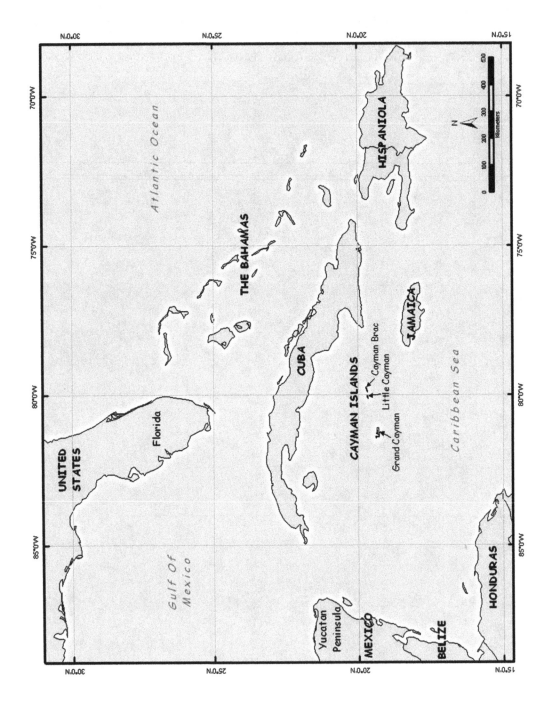

Grand Cayman

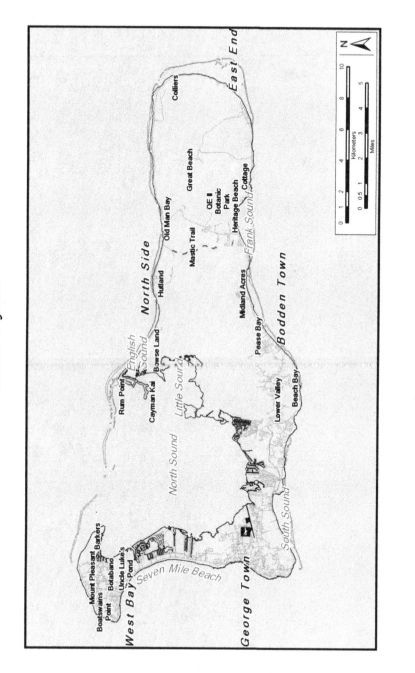

BUTTERFLIES
OF THE
CAYMAN ISLANDS

R. R. ASKEW and P. A. van B. STAFFORD

BUTTERFLIES
OF THE
CAYMAN ISLANDS

R. R. ASKEW and P. A. van B. STAFFORD

Apollo Books
2008

Technical editing: Peder Skou

Printed by: one2one a/s

Bound by: J. P. Møller Bogbinderi A/S

Published by:
Apollo Books
Kirkeby Sand 19
DK-5771 Stenstrup
Denmark
www.apollobooks.com

Sole distribution in the Cayman Islands:
National Trust for the Cayman Islands
P. O. Box 31116
558A S. Church St.
Grand Cayman KY1-1109
Cayman Islands
Phone 345-949-0121
Fax 345-949-7494
www.NationalTrust.org.ky
info@nationaltrust.org.ky

ISBN 978-87-88757-85-9

Authors' addresses:
R. R. Askew
5 Beeston Hall Mews
Beeston, Tarporley
Cheshire CW6 9TZ
United Kingdom
askew@beeston22.freeserve.co.uk

P. A. van B. Stafford
P. O. Box 1771
Grand Cayman KY1-1109
Cayman Islands
anstffrd@candw.ky

Front cover:
Cayman Lucas's Blue (*Cyclargus ammon erembis* Nabokov, 1948), photo R. R. Askew

Contents

Preface

In 1975 Dick Askew first visited the Cayman Islands as the entomologist on a Royal Society-Cayman Islands Government expedition, mounted primarily to investigate the biogeography and ecology of Little Cayman. Butterflies attracted a lot of attention, not least because a foundation for their study had previously been laid, in 1938, by an Oxford University Biological expedition. There were intriguing differences between the butterfly faunas of 1938 and 1975. Some species had declined in numbers or even disappeared, others had increased and some 'new' species had become established on the islands. In subsequent years further changes have taken place, and as the number of Cayman-based observers grows, vagrant individuals of previously unrecorded species are increasingly reported. Ann Stafford began keeping a diary of her butterfly observations in 2002, and in 2006 we decided to write this book together, putting on record our information on the changing Caymanian butterfly scene, and providing a guide book that might encourage others to get to know and help conserve the butterflies of these very attractive islands.

It would be much appreciated if unusual or interesting butterfly observations were placed on record with the National Trust for the Cayman Islands at Box 31116 Grand Cayman, Cayman Islands KY1-1205 (email: info@nationaltrust.org.ky).

Acknowledgements

We are very grateful to many people for their help at various stages in our studies. The support of the Department of Environment and of the National Trust for the Cayman Islands has been vital, and we thank in particular Mat Cottam, Gina Ebanks-Petrie and Kristan Godbeer of the former, and Lois Blumenthal, Gladys Howard, Martin Keeley, Carla Reid, Frank Roulstone and Claudette Upton of the National Trust. Previous and present staff of the Mosquito Research and Control Unit, especially Simon Conyers, John Davies, Peter Fitzgerald, Marco Giglioli, Francois Lesieur, Bill Petrie and Robin Todd, have given technical help and contributed valuable information. The maps were produced for us by Jeremy Olynik of the Department of Environment. We thank the trustees of the Natural History Museum (BMNH, London) for permission to photograph some mounted butterflies, and Blanca Huertas and Geoff Martin for their help whilst working in the museum. The following past or present Cayman Islands' residents or visitors, not mentioned above, have kindly made their observations available to us, allowed us to use their photographs (see also page 25) or helped in other ways: Denise Bodden, Roger and Mary Bumgarner, Fred Burton, Peter Davey, Manuel Dequito, Eugene Gerberg, Jennifer Godfrey, Andrew Guthrie, Brigitte Kassa, Mars van Liedfe, Lorna McGubbin, Wallace Platts, Tom Poklen, Joanne Ross, Ivalee Scott, Joan Steer, Paul Watler, Tom Watling and Jack Wilson. To all of these people we offer our sincere thanks. Finally, without the help and support of our respective spouses, Tish Askew and John Stafford, this book would never have come into being.

Introduction

Butterflies are a delightful and ever-present aspect of natural life in the Cayman Islands. Their abundance and colourful variety must surely impress even the least attuned to the natural world. They flourish in the tropical climate and rich flora of the islands, and it would be good to have confidence that this situation will continue. We cannot afford, however, to be complacent about the future. The islands are not large and butterfly populations may often be small and vulnerable to environmental change. As building development on the islands proceeds, quite rampantly on Grand Cayman, butterfly habitats are inevitably destroyed. However, the impact of development on natural life can be ameliorated by sympathetic planning, the stocking of parks and gardens with native plants and minimal use of insecticides. Butterflies are active in daylight, easy to see and mostly not difficult to identify; they are, therefore, an ideal indicator of the health of the environment.

The primary aim in writing this book is to enable residents and visitors to the Cayman Islands to identify the butterflies, both 'true' butterflies and Skippers, and thereby be in a position to add to our knowledge. There is much to be learnt. It is important to have knowledge of the status of butterfly populations over a period of years so as to be able to detect any changes and be alerted to the need for conservation measures. The butterfly fauna is dynamic, not stable, and over the short span of seventy years, since the first comprehensive survey, there have been significant changes. There have been colonizations and extinctions, as well as the not infrequent but brief appearances of vagrant species. We place on record here our information on these events so that any future changes in the butterfly fauna may be seen in an historical perspective.

The order Lepidoptera

Butterflies belong to the insect order Lepidoptera. This name means scale-winged and refers to the flattened hairs or scales which give the wings their colours. The order Lepidoptera comprises several superfamilies, but in this book we consider only two, Papilionoidea ('true' butterflies) and Hesperioidea (Skippers). All other superfamilies of Lepidoptera are made up of moths. The traditional division of Lepidoptera into butterflies (Rhopalocera) and moths (Heterocera) is convenient but has little scientific merit, some superfamilies of moths being as different from each other as they are from butterflies.

The Cayman Islands

The three Cayman Islands are Grand Cayman, the largest, and the small Sister Islands of Little Cayman and Cayman Brac which together are sometimes referred to as the Lesser Caymans. They are the projecting peaks of a submarine ridge which is a continuation of the Sierra Maestra of south-eastern Cuba. Grand Cayman, the most westerly and most populated of the three islands, lies 180 miles (240 km) south of the Isle of Pines (or Isla de la Juventud, Isle of Youth) and 161 miles (257 km) west-north-west of Jamaica in the western Caribbean. It is 60 miles (97 km) west-south-west of Little Cayman. Little Cayman is only five miles (8 km) west of Cayman Brac, which is about 119 miles (190 km) west of Cuba. All three islands are relatively narrow from north to south. Grand Cayman has a land area of 123 square miles (197 km²) but Little Cayman and Cayman Brac are each only about 30 square miles (50 km²) in area and about nine times as long (west to east) as wide.

The islands are coral limestone, of Miocene age centrally but recent peripherally. Cayman Brac rises to an altitude of 140 feet (43 m), but Grand Cayman and Little Cayman attain scarcely half this height.

There are extensive areas of Black Mangrove (*Avicennia germinans*), especially on Grand Cayman and Little Cayman, and on the former island these are criss-crossed by dykes beside which run unsurfaced roads with varied low-growing vegetation. There is dry, evergreen woodland on the central bluff limestone, and the agricultural clearance of this woodland, particularly on Grand Cayman, has created areas of rough pasture. The beach ridges have a xerophilous shrub and herbaceous plant cover, and low-lying areas with a high water-table may be colonized by *Salicornia* and *Sesuvium*. Urban development on Grand Cayman and Cayman Brac has been accompanied by the establishment of parks and gardens. All of these habitats support at least some butterfly species.

Earlier accounts of the butterflies

The first mention in the literature of a butterfly in the Cayman Islands seems to be the description of *Papilio andraemon tailori*. This is the subspecies of swallow-tail, now in the genus *Heraclides*, which is endemic to Grand Cayman, and was described in 1906 by The Hon. Walter Rothschild and Karl Jordan (1906) (full details of all cited papers are provided in the list of references). However the foundations of our knowledge of Caymanian butterflies were undoubtedly laid by C. Bernard Lewis and Gerald H. Thompson, the entomologist members of the Oxford University Biological Expedition to the Cayman Islands in 1938. They were based on Grand Cayman from 17 April to 27 August, but spent 18 to 28 May on Cayman Brac and 28 May to 10 June on Little Cayman. The butterflies collected and observed are fully discussed by Carpenter and Lewis (1943).

Between 4 July and 14 August 1975, the joint Royal Society – Cayman Islands Government Expedition spent several weeks on Little Cayman, eight days on Grand Cayman and two days on Cayman Brac. A study of the butterflies was published by Askew (1980). In August 1985, as a follow-up to the 1975 expedition, Askew spent a further 21 days on Grand Cayman and reported on apparent changes to the butterfly fauna of the island (Askew 1988); a further six visits to Grand Cayman and one each to Little Cayman and Cayman Brac, mostly between January and March, have followed.

Schwartz *et al.* (1987) give details of the more notable discoveries made by Albert Schwartz and Fernando Gonzalez between 27 November and 1 December 1985, and Miller & Steinhauser (1992) record their findings from a one-week collecting trip at the beginning of November 1990. Both of these studies included observations made on both Grand Cayman and Cayman Brac.

The Cayman Islands' butterfly fauna is

sometimes referred to in books which cover the West Indies as a whole. Of these, Riley (1975), and the very comprehensive book by Smith, Miller & Miller (1994), include accounts, illustrated by paintings, of all Caymanian species, and Stiling (1999) presents photographs of several butterfly species found in the Islands. Richard Ground (1989), in his book *Creator's Glory* which is devoted to photographs of wildlife in the Cayman Islands, depicts a number of butterflies. Books by Brown & Heineman (1972) and by Garraway & Bailey (2005) on the butterflies of Jamaica, by Hernández (2004) on the butterflies of Cuba, by DeVries (1987) on part of the Costa Rican butterfly fauna and by Gerberg & Arnett (1989) on butterflies of Florida, also include much information relevant to Cayman butterflies.

Status of butterfly species on the three Cayman Islands

The founding study of Cayman butterflies in 1938 (Carpenter & Lewis 1943) gives us a good indication of the numerical status of the species on each of the Cayman Islands at that time, but the next comprehensive survey was not made until 1975. Subse-

quent surveys have mostly concentrated upon Grand Cayman, and knowledge of the status of butterfly species since 1975 on Cayman Brac, and especially Little Cayman, is rather fragmentary, although both were visited by Askew in 2008. Grand Cayman butterflies were surveyed in 1985 and 1995, and thereafter with increased frequency. A growing number of butterfly enthusiasts resident in the Cayman Islands should ensure that a more complete history of the Islands' butterflies is compiled during the twenty-first century.

Check-lists of butterfly species on each of the Cayman Islands, compiled from all available sources, are presented below. Names used are as in Lamas (2004). An estimate of the numerical status of each species on a scale of 0 to 4, as assessed on each survey or year(s), is included.

0 - Not recorded.
1 - Very rare. Only one or two individuals seen in total at one or two sites.
2 - Uncommon and local. More than two individuals seen at one site, or three to five individuals seen at two or more sites.
3 - Quite common. Six to about twelve individuals seen at two or more sites.
4 - Very common and widespread.

Check-list of butterflies of Grand Cayman

	1938	'75	'85	'90 -'95	'96 -'98	2000 -'02	'03 -'04	'05 -'06	'07 -'08
DANAIDAE									
Danaus plexippus	2	1	1	1	0	1	2	3	3
D. gilippus	4	3	4	3	4	4	3	4	3
D. eresimus	4	2	4	2	4	4	4	4	4
NYMPHALIDAE									
Anaea troglodyta	0	0	2	0	3	3	3	4	4

	1938	'75	'85	'90 -'95	'96 -'98	2000 -'02	'03 -'04	'05 -'06	'07 -'08
Memphis verticordia	4	1	2	1	2	2	2	3	3
Marpesia eleuchea	0	0	0	0	0	2	0	0	0
M. chiron	0	0	0	0	0	0	0	1	0
Hamadryas amphichloe	0	0	0	0	0	0	0	1	0
Hypolimnas misippus	0	1	0	0	0	0	0	0	0
Junonia evarete	3	4	4	4	4	4	4	4	4
J. genoveva	4	0	1	4	3	3	0	4	2
Anartia jatrophae	4	4	4	4	4	4	4	4	4
Siproeta stelenes	2	1	1	1	2	2	2	3	2
Phyciodes phaon	4	4	4	4	4	4	4	4	4
Vanessa cardui	0	0	1	0	0	1	0	0	0
Euptoieta hegesia	4	1	1	1	0	4	4	4	4
HELICONIIDAE									
Agraulis vanillae	4	4	4	4	4	4	4	4	4
Dryas iulia	3	1	3	3	4	4	4	4	4
Heliconius charithonia	4	1	3	3	3	2	3	2	3
LYCAENIDAE									
Chlorostrymon maesites	0	0	0	0	0	0	1	0	0
Strymon martialis	0	0	1	2	0	1	1	1	1
S. acis	3	0	1	2	2	0	0	1	0
S. istapa	4	3	3	0	4	4	4	4	3
Electrostrymon angelia	0	0	1	0	0	0	0	2	0
Brephidium exilis	2	0	2	2	0	2	0	3	3
Leptotes cassius	3	3	0	2	3	2	3	1	3
Hemiargus hanno	4	2	4	2	3	3	3	4	3
Cyclargus ammon	4	4	4	2	3	4	2	3	4
PIERIDAE									
Glutophrissa drusilla	0	1	0	0	0	0	1	0	0
Ascia monuste	4	3	4	4	4	4	4	4	4
Eurema daira	0	0	1	2	0	1	0	0	0
E. elathea	4	3	3	4	4	4	4	4	3
Pyrisitia messalina	4	0	0	0	0	0	0	0	0
P. lisa	4	4	4	2	2	2	0	0	0
Abaeis nicippe	3	2	0	1	0	2	0	0	0
Nathalis iole	0	0	1	0	0	0	0	0	0
Anteos maerula	0	0	1	1	0	2	0	0	0
Phoebis sennae	4	3	3	4	3	4	4	4	4
P. agarithe	0	0	3	4	3	3	0	3	3
P. philea	0	0	0	0	0	1	2	0	1
Aphrissa statira	1	0	0	1	0	0	0	3	0
A. orbis	0	0	0	0	1	0	1	0	0

	1938	'75	'85	'90 -'95	'96 -'98	2000 -'02	'03 -'04	'05 -'06	'07 -'08
PAPILIONIDAE									
Battus polydamas	2	0	0	2	0	0	0	0	3
Heraclides andraemon	4	3	3	3	3	4	4	4	4
HESPERIIDAE									
Urbanus proteus	4	2	3	3	3	2	2	4	4
U. dorantes	0	0	1	1	0	0	0	0	0
Cymaenes tripunctus	4	3	3	2	2	1	0	3	3
Hylephila phyleus	4	0	3	3	2	2	2	3	2
Asbolis capucinus	0	0	0	0	0	1	2	2	2
Calpodes ethlius	0	1	0	0	0	2	0	1	2
Panoquina lucas	4	1	3	3	0	2	1	4	2
P. panoquinoides	4	3	0	0	0	0	0	0	0
Total species 52 Totals	33	29	36	35	27	38	30	36	33

[1985 data includes that of Schwartz *et al.* (1987); 1990-1995 data includes that of Miller & Steinhauser (1992) for 1990]

Check-list of butterflies of Little Cayman

	1938	'75	2007 -'08
DANAIDAE			
Danaus plexippus	0	0	1
Danaus gilippus	3	2	0
Danaus eresimus	0	0	2
NYMPHALIDAE			
Memphis verticordia	3	4	4
Junonia evarete	2	0	0
J. genoveva	0	4	4
Siproeta stelenes	0	0	2
Phyciodes phaon	0	0	2
Euptoieta hegesia	4	4	4
HELICONIIDAE			
Agraulis vanillae	4	4	4
Heliconius charithonia	0	3	4
LYCAENIDAE			
Strymon martialis	1	3	3
S. acis	1	3	4
S. istapa	0	2	4
Leptotes cassius	3	3	4

	1938	'75	2007 -'08
Hemiargus hanno	0	2	4
Cyclargus ammon	2	4	4
PIERIDAE			
Glutophrissa drusilla	2	4	4
Ascia monuste	3	4	4
Eurema daira	0	2	0
Eurema elathea	0	0	4
Pyrisitia lisa	0	3	2
Abaeis nicippe	0	3	0
Nathalis iole	0	1	0
Phoebis sennae	3	4	3
PAPILIONIDAE			
Heraclides andraemon	4	3	3
H. aristodemus	3	2	1
HESPERIIDAE			
Phocides pigmalion	3	2	4
Hylephila phyleus	0	0	3
Atalopedes mesogramma	0	0	1
Panoquina panoquinoides	2	3	2
Total species 31 Totals	16	23	26

Check-list of butterflies of Cayman Brac

	'38	'75	'85-'90	'94	'07 -'08
DANAIDAE					
Danaus plexippus	2	0	0		3
D. gilippus	3	0	0		0
NYMPHALIDAE					
Memphis verticordia	3	1	1		0
Junonia genoveva	3	2	3		4
Anartia jatrophae	1	1	0		3
Siproeta stelenes	0	0	0		2
Phyciodes phaon	0	0	0		3
Euptoieta hegesia	4	1	4		4
HELICONIIDAE					
Agraulis vanillae	4	4	4		4
Dryas iulia	0	0	1		0
Heliconius charithonia	0	1	2		2

	'38	'75	'85-'90	'94	'07-'08
LYCAENIDAE					
Eumaeus atala	0	0	4	+	3
Strymon acis	4	2	1	+	1
S. istapa	0	0	1	+	4
Electrostrymon angelia	0	0	1	+	0
Leptotes cassius	3	2	3		4
Hemiargus hanno	0	0	3	+	4
Cyclargus ammon	2	3	4		3
PIERIDAE					
Glutophrissa drusilla	2	1	3		4
Ascia monuste	4	3	3		4
Eurema daira	0	2	3		0
E. elathea	0	0	2		4
Pyrisitia lisa	0	3	0		1
P. nise	0	0	1		0
Abaeis nicippe	3	2	0	+	0
Phoebis sennae	4	3	1		4
P. agarithe	0	0	1		2
Aphrissa orbis	0	0	0	*	0
PAPILIONIDAE					
Heraclides andraemon	4	3	3		3
HESPERIIDAE					
Urbanus proteus	2	0	3		1
Cymaenes tripunctus	0	3	2		0
Hylephila phyleus	3	0	0		4
Atalopedes mesogramma	0	0	0		2
Panoquina panoquinoides	3	3	0		2
Total species 34					
Totals	18	18	23		25

[Records under 1985 - 1990 are from Schwartz *et al.* (1987) and Miller & Steinhauser (1992); under 1994, + indicates a specimen collected by P. Davey and housed in the National Trust, and * is a record in Smith *et al.* (1994)]

Changes over time

A glance at the check-lists suggests that there have been qualitative and quantitative changes in the butterfly faunas of each of the three islands over quite short periods. It must be emphasized, however, that our assessments of status are made with many variables, such as time of year, weather conditions and duration of recording. Also, in the absence of continuous observation, it is impossible to know whether a species not seen in a particular year is really absent or has simply been overlooked. It was only on Askew's fifth visit to Grand Cayman that the Pygmy Blue (*Brephidium exilis*) was located, despite being intensively sought on previous occasions. The likelihood is that it was present but, being a tiny brownish butterfly, had remained undiscovered. On the other hand, the sudden appearance on Grand Cayman in 1985 of the conspicuous Cuban Red Leaf Butterfly (*Anaea troglodyta*) and Large Orange Sulphur (*Phoebis agarithe*) were almost certainly recent colonizations.

Accepting the limitations of our records, however, those for Grand Cayman, which are the most complete, suggest that in the period 1938 to 2008, 15 of the 52 recorded species are vagrants or temporary, short term colonists, 27 are permanent, breeding residents, six species (*Anaea troglodyta, Phoebis agarithe, P. philea, Battus polydamas, Asbolis capucinus* and *Calpodes ethlius*) appear to have become established on the island, but four species (*Pyrisitia messalina, P. lisa, Abaeis nicippe, Panoquina panoquinoides*) present in 1938 may now be extinct. This represents a 16 percent ((6 + 4) 100/ (27 + 6) + (27 + 4)) change in the butterfly fauna, rather less than the 20 percent change estimated for the

15

period 1938 to 1985 (Askew 1988). Almost thirty percent of the total of species recorded for Grand Cayman can be classified as either vagrants (e.g. *Hamadryas amphichloe, Marpesia chiron*), or temporary colonists breeding for only a short period (e.g. *Marpesia eleuchea, Aphrissa statira*). Quite possibly, some of these species may in time become more permanently established, as *Anaea troglodyta, Phoebis agarithe* and *Asbolis capucinus* have done, but it is clear that Grand Cayman receives a steady influx of vagrant species. This helps to drive faunal change. Numbers of colonizing species and of those becoming extinct should be about equal, so that the size of the Grand Cayman butterfly fauna at any one time should remain around 34 to 39 species, just above the range recorded in the eight survey periods.

Size of the butterfly fauna

An overall total of 57 butterfly species have been reliably recorded from the Cayman Islands, 52 from Grand Cayman, 31 from Little Cayman and 34 from Cayman Brac. Little Cayman is under-recorded. Since 1984 the total for Grand Cayman has risen by 16, for Little Cayman by 7 and for Cayman Brac by 12.

The number of species that an island can support is strongly correlated with its size. Large islands encompass greater environmental heterogeneity than small islands, providing more ecological niches and therefore able to support a larger fauna. When faunas of islands of different sizes, but of similar location, climate and topography, are compared, a remarkably close positive correlation between land areas and species numbers in a taxon (e.g.

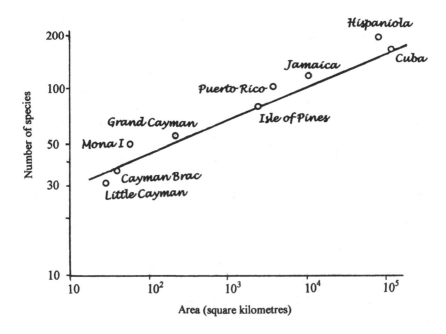

Figure 1. Relationship between numbers of butterfly species and land area for the Cayman Islands and other west Caribbean islands. Scales on both axes are logarithmic.

reptiles, birds, butterflies) may be found. This species-area relationship applies very well to butterflies on islands of the northern Caribbean (Askew 1988, 1994); a regression of species numbers against land area on a logarithmic scale for Cuba, Isle of Pines, Hispaniola, Puerto Rico, Mona, Jamaica and the three Cayman Islands (Figure 1) has a correlation coefficient of 92 percent. The butterfly fauna of Grand Cayman is slightly larger than that suggested by the species-area regression, whilst that of relatively little studied Little Cayman is smaller. Mona Island, which in size is between Cayman Brac and Grand Cayman, has a very large recorded fauna, but this includes a substantial proportion of vagrants and temporary colonists. The correlation coefficient would be raised if only regularly occurring breeding species were considered to make up the number of butterfly species on each island.

Affinities of the butterfly fauna

Only five subspecies of butterfly are endemic to (i.e. evolved in) the Cayman Islands. These are *Memphis verticordia danielana* (Witt, 1972), *Dryas iulia zoe* Miller & Steinhauser, 1992, *Brephidium exilis thompsoni* Carpenter & Lewis, 1943, *Cyclargus ammon erembis* Nabokov, 1948 and *Heraclides andraemon tailori* (Rothschild & Jordan, 1906) on Grand Cayman. *M. v. danielana* and *C. a. erembis* fly on all three islands, *D. i. zoe* on Grand Cayman and Cayman Brac, but *B. e. thompsoni* and *H. a. tailori* are restricted to Grand Cayman. No full species is certainly endemic although it is uncertain whether or not *C. a. erembis* should be accorded full specific status.

The islands are, it seems, insufficiently isolated to permit the extensive evolution of locally adapted forms. Butterflies in general are highly mobile, and the expanse of two hundred kilometres of Caribbean Sea separating the islands from Cuba (and also Jamaica) has not prevented frequent immigrations, no doubt assisted by the often strong prevailing easterly to north-easterly winds. There must also have been accidental introductions through human agency, accounting, for example, for at least some of the appearances of the Canna Skipper (*Calpodes ethlius*) on Grand Cayman.

Clench (1964) and Scott (1972) concluded that the Cayman butterfly fauna is derived predominantly from Cuba, and this view is supported by our own data and that of Smith *et al.* (1994) which show that all 57 Caymanian species are included in the much larger fauna of Cuba, with 46 in Jamaica, 45 in both southern Florida and Hispaniola, 39 in Puerto Rico and 32 in Mona Island.

There is clearly a great deal of butterfly movement between islands, working against the evolution of endemic forms, but on the other hand the fauna of each of the three Cayman Islands has its own peculiarities. The large proportion of vagrant species recorded from Grand Cayman but not from Little Cayman or Cayman Brac is chiefly a consequence of much more intensive observation on the largest island, but there are other faunistic differences between the islands, both qualitative and quantitative, that are probably attributable to ecological factors. *Heraclides andraemon* occurs as different subspecies on Grand Cayman and on the two smaller Cayman Islands, *Eumaeus atala* (and perhaps *Pyrisitia nise*) may be resident only on Cayman Brac, *Heraclides aristodemus* and *Phocides pigmalion* are known only from Little Cayman, and *Atalopedes mesogramma* has so far been found only on the Sister Islands. *Brephidium exilis* has been long established on Grand Cayman but has never been observed on the Sister Islands. Several instances of quantitative differences in the faunas of the three islands may be found in a comparison of the check-lists; for example, *Glutophrissa drusilla* is much commoner on both Little Cayman and Cayman Brac than it is on Grand Cayman, whilst *Eurema elathea* and *Phyciodes phaon*, both very common on Grand Cayman, have appeared only relatively recently on the Sister Islands.

Hurricane impact on Grand Cayman butterflies

It is tempting to attribute much of the dynamism of the butterfly faunas of the Cayman Islands to the frequent high winds of the region. Although hard evidence of wind-assisted inter-island butterfly movement is lacking, there is no shortage of circumstantial evidence. For instance, *Greta* (probably *cubana*) (page 131) was in all probability blown to Grand Cayman in 2005 by Hurricane Dennis. The occasional records in 2005 of *Hamadryas amphichloe* (page 44) and *Marpesia chiron* (page 42) could also have been of wind-blown insects; 2005 was a year of especially frequent high winds and storms.

The most severe hurricane in recent years was Hurricane Ivan which struck Grand Cayman on 11 and 12 September, 2004. Ivan had a profound affect on wildlife. Mature trees were blown over and large areas were left inundated by sea water after the storm surge. A lot of vegetation was destroyed, but the ground was opened up to colonization by invasive plants. Birds, particularly the smaller passerines such as Bananaquits, Flycatchers and the Caribbean Elaena, almost disappeared, and populations eighteen months later were still well below pre-Ivan levels.

Immediately after Hurricane Ivan, adult butterflies were very scarce on Grand Cayman, but their populations quickly recovered, immature stages presumably having survived much better than the adult insects. In George Town in the second half of September 2004, *Anartia jatrophae* and *Agraulis vanillae* were the first butterflies to reappear after the hurricane, followed in October by *Junonia* species, *Phyciodes phaon*, *Euptoieta hegesia*, *Eurema elathea*, *Phoebis sennae*, *Hylephila phyleus* and *Panoquina lucas*. In November, all of the commoner resident species could be seen in plenty. This strong recovery of butterfly populations has been attributed to the decimation of insectivorous birds. This may well have had a considerable affect, but it can scarcely account for the fast recovery of unpalatable butterflies such as Danaidae. Another aftermath of Hurricane Ivan was the luxuriant growth of herbaceous plants on cleared ground, and vigorous shooting from broken shrubs and trees. Almost all of Grand Cayman's resident butterflies could thus have benefited.

Where to see butterflies

Butterflies may be seen virtually anywhere in the Cayman Islands, even in the centre of George Town. They are most numerous where there is a plentiful variety of herbaceous flowers in sunny, open situations such as the landward side of beach ridges, wasteland and disturbed ground with secondary vegetation, but the more specialized species will mostly be found along wood edges and in light woodland, including some parks and large gardens.

The parks on Grand Cayman, established by the Dart Foundation in partnership with the Cayman Island Government, are usually especially rich in butterflies. These parks have been carefully landscaped and planted with a good proportion of native plants. The Dart Family Park in George Town should well repay a visit. The Boatswain's Beach Adventure Park, developed from the old Turtle Farm at West Bay, includes a nature trail winding through Caymanian dry woodland with a butterfly grove where again many of Grand Cayman's butterflies may be seen in the wild.

Different butterfly species have different habitat requirements, although their mobility enables them to often seek nectar quite far from the food-plants that nourish their larvae. Parks and gardens are good places to see a range of species, but so also are overgrown roadside verges and uncultivated land where a partly differing range may be flying. On the other hand, the only place to look for Pygmy Blues (*Brephidium exilis*) is in patches of the succulent Sea-pusley (*Sesuvium*) amongst which Glasswort (*Salicornia*), the larval food-plant, is growing.

Counts were made of the numbers of butterflies seen whilst walking slowly for a known time in each of various habitats in Grand Cayman and Little Cayman (Table 1). Although influenced by several uncontrolled variables, this method provides an adequate general pattern of relative butterfly abundance and distribution. It is a variant of the widely-used Pollard Walk (Pollard 1977) in which butterflies are counted on a fixed transect, a technique

which could easily and very usefully be applied in the Cayman Islands to monitor seasonal and yearly changes.

Table 1. Butterflies, with numbers seen per minute in brackets, in various habitats on Little Cayman (L) and Grand Cayman (G). Butterflies are identified only to genus, and only the three most numerous taxa in each habitat are listed. Data for Little Cayman were obtained in 1975 (Askew 1980).

Beach ridge
L *Agraulis* (0.7), *Euptoieta* (0.5), *Glutophrissa* (*Appias*) (0.2)
G *Hemiargus/Cyclargus* (0.3), *Junonia* (0.3), *Agraulis* (0.2)

Dyke road
G *Junonia* (1.7), *Agraulis* (0.5), *Anartia* (0.4)

Mangrove
L *Memphis* (<0.1), *Euptoieta* (<0.1), *Glutophrissa* (<0.1)
G *Junonia* (0.1), *Ascia* (<0.1)

Pasture
G *Eurema/Pyrisitia* (0.4), *Agraulis* (0.3), *Phyciodes* (0.3)

Uncultivated land
G *Junonia* (1.5), *Anartia* (0.2), *Ascia* (0.2)

Ironshore woodland
L *Glutophrissa* (0.5), *Memphis* (0.3), *Heraclides* (0.2)
G *Junonia* (0.5), *Agraulis* (0.4), *Anartia* (0.2)

Bluff woodland
L *Glutophrissa* (0.3), *Agraulis* (0.3), *Memphis* (<0.1)
G *Ascia* (0.7), *Agraulis* (0.4), *Eurema/Pyrisitia* (0.2)

A reference collection of Caymanian butterflies is maintained in the office of the National Trust for the Cayman Islands and may be looked at by anyone wishing to examine or compare named specimens.

The Butterfly Farm in George Town, Grand Cayman provides an opportunity to view, under captive conditions, some of the more spectacular tropical butterflies, their caterpillars and pupae. Species that occur naturally in the Cayman Islands are not represented in the cages. This safeguards the genetic integrity of Cayman populations in the event of escapes from the Butterfly Farm. However free-flying butterflies are encouraged to frequent the vicinity of the Butterfly Farm by the cultivation of native larval food-plants and nectar sources.

When to see butterflies

Butterflies can be seen all year round in the Cayman Islands but are especially numerous between September and April. The larval food-plants seem always to be available, and there is clearly much less seasonality in butterfly life cycles in the Cayman Islands compared to more temperate latitudes. As a broad generalization supported by very little information, it appears that in most Cayman butterflies there are probably overlapping generations with all life stages, egg, larva, pupa and adult, present at all times. This, however, is a subject requiring study, and the Pollard walk, mentioned above, would be a simple method of providing data on the seasonal variation in numbers of adult butterflies.

The butterfly life cycle

Insects are invertebrate animals, lacking an internal skeleton but instead having, for body support and muscle attachment, a hard (or at least reasonably firm) exoskeleton or integument on the outside of the body. The bulk of the integument comprises the cuticle which is composed of layers, some hard and supporting, some with water-proofing qualities which impede water loss through the body surface. The rigidity of the cuticle requires that the insect's body is jointed to permit movement, a feature of all arthropods (insects, spiders, crabs and the like), and that it is periodically cast off and replaced by a new, larger cuticle to accommodate the body's growth. Shedding or moulting of the cuticle gives the insect an opportunity to fundamentally change or metamorphose its body form during development. Once the adult stage has been reached, there is no further moulting or growth.

In so-called less specialized insects, such as dragonflies, cockroaches, crickets and true bugs, the creature that hatches from the egg, the larva or nymph, is generally very active with well-developed compound eyes, and it progresses to the adult stage by a series of moults or ecdyses with a step-wise increase in size at each but no radical transformation in external appearance. This is termed incomplete metamorphosis. The larval stages are unable to fly, but they have externally developing wings visible on the thorax as small wing-buds, and because of this insects with incomplete metamorphosis are described as exopterygote insects.

In contrast, a complete metamorphosis is characteristic of all 'higher' insects such as butterflies and moths, beetles, true flies, wasps and ants. In these the larval stages (commonly known as caterpillars, grubs or maggots) tend to be relatively slow-moving creatures, living surrounded by their food, eating voraciously and growing rapidly. They do not have compound eyes. There is no external trace of wings, but these are developing inside the thorax; butterflies and other insects with a complete metamorphosis are described as endopterygotes. Between the final larval stage and the adult, the endopterygote insect interposes another distinct stage, the pupa. The pupal stage is needed for reorganization of the body, for switching between the divergent larval and adult body forms. It is an almost immobile stage, but one in which there is intense physiological activity, the larva's organs being restructured or replaced by adult organs. At the final larval moult, the wings are everted and are visible on the surface of the pupa as wing cases; other structures of the adult such as its legs, mouthparts, antennae and compound eyes are similarly visible on the pupa. There is no feeding or growth during the pupal stage.

A butterfly commences life as an egg laid by its mother, singly or in a clutch, on a suitable food-plant. A secretion produced by the female butterfly glues the egg to the plant. Butterfly eggs are variously spherical to lemon- or spindle-shaped with flat bases, and the egg shell or chorion may be smooth but is more often quite ornately sculptured.

After a time, usually just a few days in tropical temperatures, embryonic development is completed and the first stage larva or caterpillar chews its way out of the egg. Its first meal is often the egg shell (in some Papilionidae it is a waxy covering to the egg) and this sustains it until it finds a plant meal. The mouthparts of the larva consist principally of biting jaws (mandibles), quite unlike the sucking proboscis of the adult butterfly. Openings to the silk-producing salivary glands are adjacent to the bases of the mandibles. Behind the head, the three-segmented thorax is differentiated from the rest of the body only by having a pair of short, jointed legs on

each of its segments. An abdomen of ten segments completes the body with a pair of fleshy, unjointed false legs (prolegs) on abdominal segments three to six and a fifth pair on the last body segment. The prolegs are equipped with small hooks which enable the caterpillar to firmly cling to its food-plant. Small round or oval openings, the spiracles, are visible on the sides of some of the thoracic and abdominal segments, one pair to a segment, and it is through the spiracles that respiratory gaseous exchange takes place between the internal tracheal tubes and the atmosphere. The caterpillar eats and grows, and moults usually five times. The stages between moults can be referred to as instars. After the fifth moult the pupa or chrysalis is usually revealed. Prior to this final larval moult, the caterpillar stops feeding and sometimes walks away from its food-plant to find a suitable pupation site. Here it spins a pad of silk, to which it attaches itself by the terminal pair of prolegs (the cremaster), and sometimes also a silken girdle over the thorax which gives additional support.

The butterfly pupa, often called a chrysalis because it may be adorned with golden markings, is static and immobile except for a capacity to twitch or wiggle its abdomen. It is held in position by its cremaster, and in Papilionidae and Pieridae by a silken girdle. The cuticle is moulded to the form of the adult butterfly with head, thorax and abdomen clearly differentiated, and the positions of compound eyes, antennae, proboscis, legs and wings indicated by cases. The butterfly forms beneath the pupal cuticle, and a day or so before it emerges the wing pattern becomes visible through the pupal wing case. Emergence or eclosion involves an increase in pressure inside the front of the body, resulting

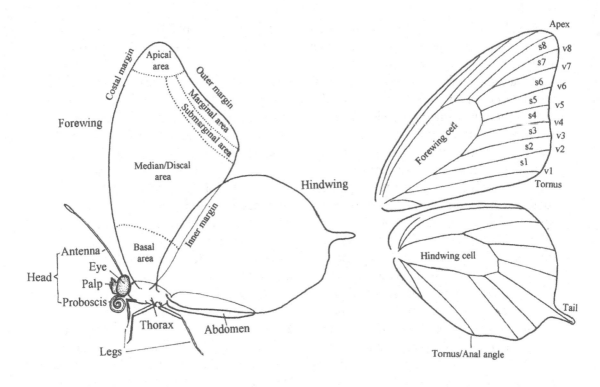

Figure 2. Morphology of the adult butterfly illustrating features mentioned in the text. Wing veins (v) and spaces (s) are numbered on the right hand drawing of a forewing.

from contraction of muscles more posteriorly, and this causes the pupal cuticle of the head and thorax to split. The butterfly crawls clear of the pupal case and hangs, wings down, from either the empty case or adjacent vegetation. On first emergence, the wings are small and crumpled, but within minutes they start to expand as body fluid (haemolymph) is pumped into their veins by muscular contractions of the body. Two or three drops of a reddish brown waste fluid (the meconium) are voided through the anus, and the newly emerged butterfly hangs motionless for a period of an hour or more whilst its wings dry and harden, before taking its maiden flight. Female butterflies mate very early in their lives, in Heliconiidae sometimes even before they have fully emerged from the pupal case.

Butterfly morphology

Those parts of the adult butterfly's body which are mentioned in the species' descriptions may be located by reference to Figure 2. Particular attention is paid to the wings because these organs provide the majority of characters that enable a species to be identified. The male genitalia are also rich in diagnostic characters, but

Cayman butterflies should be identifiable without having to resort to examining the sex organs.

The wings are essentially three-sided and each edge or margin is named (costal, outer and inner). Areas of the wing surface, rather arbitrarily delimited, are a basal area adjacent to the wing attachment to the thorax, apical and tornal areas respectively at the anterior (apex) and posterior (tornus) outer angles. A marginal area runs along the outer margin with a submarginal area basal to it. The central area of the wing is referred to as the discal or median area, sometimes subdivided into a basal submedian area, median area and distal postmedian area. The wings are strengthened by the so-called veins. The arrangement of veins (venation) is based on a plan common to all butterflies, indeed to all insects. The veins are all named, but so far as butterflies are concerned it is usual to refer to them by numbers. On each wing the veins that extend to the outer margin are numbered from the tornus or anal angle forwards to the wing apex. The spaces between veins are also numbered, space 1 being anterior to vein 1, space 2 anterior to vein 2, and so on. An important feature is an area termed the cell which extends across the basal half or more of each wing, and is defined by veins from which the numbered veins radiate.

The Butterflies

On the following pages each species of butterfly that, to our certain knowledge, has been recorded alive in the wild in the Cayman Islands is considered separately. The species are grouped in their families, the characteristics of each of which are described. The families in both superfamily Hesperioidea (Skippers) and superfamily Papilionoidea ('true' butterflies) are considered. All other superfamilies of Lepidoptera comprise the moths. The scientific names of butterflies follow those in Lamas (2004). Accounts of species follow a common plan and are arranged in sections as follows.

Names

The English common or vernacular name (or names) of the butterfly species heads each account. For most people the common name is more easily remembered and pronounced than the scientific name, but no international rules or code govern its application so that it may be only locally recognized. Furthermore, it conveys, at best, only very limited information on the relationships of the butterfly. Below the common name we give the scientific name. It is written in italics and is in two parts, the genus commencing with a capital letter, and the specific (or trivial) name which always begins with a lower case letter. The scientific name is followed by the name of the person who first published it with a description of the butterfly, and the year of publication. Name and year are sometimes enclosed in brackets; this is the convention for showing that the species was originally described in a genus different to the one in which it is now

placed. The scientific name is unique and universally recognized.

Recognition

The six plates include photographs of one or more set specimens of each Cayman butterfly species, and the location of these images is given below the scientific name of each species. The section commences with the range of forewing lengths (FWL) in millimetres, measured where possible on Caymanian specimens. The measure is taken from the base of the forewing, at its articulation with the thorax, to the wing apex. It is to be taken only as a guide to size and particularly small individuals outside the given range may sometimes be found. In the majority of large or medium-sized butterflies, females tend to be a little larger than males.

An account of the wing pattern describes characteristic features of the insect, emphasizing how it may be distinguished from similar species. The various terms used to describe different regions of the wings are shown in Figure 2.

Subspecies

Many butterfly species vary geographically, populations from different places often appearing somewhat different. This applies particularly to island populations. Where movement between islands is limited, locally adapted forms are able to evolve. When a local population is recognizably distinct it may be given a subspecific name. This is added to the generic and specific names to form a trinomial. The application of subspecific epithets tends to be rather subjective and imprecise, and they have, perhaps, been overused in the past. However, the subspecies can be a useful concept and we employ it here where we believe it to be justified. There are five subspecies (page 17) in the Cayman butterfly fauna which are considered to be endemic, having evolved in the Cayman Islands, and this provides a measure of the extent of isolation of the fauna.

Species' range

This is a description, in rather broad terms, of the geographical area in which the species (not just the subspecies) is found.

Cayman Islands distribution

In this section we list the islands in the Cayman group from which the species has been recorded.

Habitat

A brief indication of the type of situation in which the butterfly is most likely to be seen is given here. Butterflies however, especially the larger species, fly around a great deal, and may be encountered almost anywhere on these relatively small islands.

History

We provide here an account of our knowledge of each species as a Caymanian insect, from 1938 when the field-work for the base-line study of Carpenter & Lewis (1943) was done, to the present time.

Biology

This section includes short accounts of the immature stages in the life history (egg, larva or caterpillar, pupa or chrysalis), the larval food-plants and adult nectar plants, and adult biology including any characteristic features of flight or other behaviour. Both scientific and vernacular names of the larval food-plants, but only the scientific names of nectar plants, are given in the text. Scientific and vernacular names of all plants mentioned are listed in a later section (page 148). The immature stages have been seen for only a small proportion of Caymanian butterflies and descriptions for most species are taken from the literature. There is a good opportunity here for very worthwhile contributions to be made to our knowledge by the observation and rearing of eggs, larvae and pupae.

Photographs

The photographs are of living Cayman butterflies and their immature stages, where available, or of dead specimens found in the Cayman Islands. The legend to each photograph indicates the island on which the photograph was taken (GC – Grand Cayman, LC – Little Cayman, CB – Cayman Brac), the date and the photographer's initials:

MLA – Tish Askew; RRA – Dick Askew; LB – Lois Blumenthal (National Trust GC); DB – Denise Bodden (National Trust GC); PD – Peter Davey; KDG – Kristan Godbeer (Dept of Environment); JG – Jennifer Godfrey; WP – Wallace Platts (National Trust CB); TP – Tom Poklen (Chicago); JR – Joanne Ross; FR – Frank Roulstone (National Trust GC); AS – Ann Stafford

Milkweed Butterflies

Danaidae

These large and relatively slow-flying butterflies share with the families Nymphalidae and Satyridae the feature of having reduced front legs which cannot be used for walking, and many lepidopterists regard them as a subfamily of Nymphalidae. However, unlike nymphalids, male danaids have extrusible hair pencils near the end of the abdomen which, in nearly all species, are everted as the male flutters over the female during courtship. The hair pencils have previously been impregnated with scent scales picked up from small pouches or sex brands near the centres of the male's hindwings. The scent scales contain a pheromone, an aphrodisiac, and are transferred to the female's antennae when the male scatters a cloud of them over his partner during his hovering courtship flight. The sex brands (also known as androconial patches) are visible as black spots on vein 2 on both upper and lower wing surfaces, and provide an easy way to distinguish the sexes of Cayman Danaidae. The pheromone is derived from pyrrolizidine alkaloids which are obtained by the male butterfly on feeding at certain plants, particularly Boraginaceae (Edgar 1982).

Danaid larvae lack spines on the body, a feature of larvae of Nymphalidae, but they have two or three pairs of tubercular processes which can be twitched when the larva is touched. The larvae feed principally on Apocynaceae (subfamily Asclepiadoidae). These plants are rich in cardiac glycosides, which are heart poisons to warm-blooded vertebrates. They are accumulated by the feeding larva and passed on to the adult butterfly. Both larva and adult are rendered distasteful by the noxious chemicals, and they warn off would-be predators by displaying striking and memorable colour patterns which a naïve predator with colour vision, such as a bird or lizard, will quickly learn to associate with an unpleasant experience. The education of a predator requires that a few butterflies are tasted, but danaids are able to survive a moderate amount of physical mauling by virtue of their tough, leathery integument. When attacked, a danaid will frequently feign death, and this immobility probably reduces the likelihood of serious injury.

Butterflies in general, not only Danaidae, which sequester noxious chemicals depend on their wing patterns being memorable. To this end the patterns are usually simple, in contrasting colours, repeated on upper and under wing surfaces, and similar in both sexes. The butterflies fly relatively slowly to advertise their warning livery, and sometimes come together in numbers so as to enhance their visual impact.

In order to share the burden of educating naïve predators, two or more species of butterflies may develop a similar wing pattern so that a predator that has tasted one species will generalize the unpleasant experience and, in future, avoid all species of similar appearance. In this way, predator attacks and possible

Male Monarch on Red Top (Asclepias curassavica), CB (1.ii.2008), MLA

Monarch

Danaus plexippus (Linnaeus, 1758)
Plate I (1,2)

Recognition

FWL 45-55 mm. Its size and brownish orange wings with black borders, and veins marked with black scales, distinguish the Monarch from all other Cayman butterflies. It is larger and more brightly coloured than the other two species of *Danaus* that occur in the islands, and the apical one-third of the forewing is black with pale spots. The hindwing under surface is buff coloured with black veins, and the male sex brand, a black spot against vein 2 near the centre of the hindwing, is small. There are small white spots on the top of the head and thorax, as in other species of *Danaus*.

mortalities are shared between all butterfly species displaying a similar pattern. This is known as Müllerian mimicry and a possible example of it in the Cayman Islands is the resemblance between the Queen and Soldier (below).

Another type of mimicry is known in the Monarch. In parts of its range some caterpillars feed on plants such as Dogbane (*Apocynum*) which may contain relatively low levels of toxin. The resulting butterflies are edible, but they are avoided by a predator that has had previous experience of a distasteful Monarch. This is automimicry, the edible mimic and its distasteful model both belonging to the same species, and in principle it is the same as Batesian mimicry in which mimic and model are of different species.

Female Monarch, GC (25.i.2006), MLA

Subspecies

Both the migratory subspecies of North America, *Danaus plexippus plexippus*, and a non-migratory Caribbean subspecies, *D. p. megalippe* (Hübner, 1826) have been found in the Cayman Islands. *D. p. megalippe* may be distinguished from the nominate subspecies by its less concave outer forewing margin and reduction in number and size of the whitish spots in the black marginal band on the hindwing upper surface. *D. p. plexippus* usually has two complete rows of whitish spots whereas in *D. p. megalippe* both rows, but especially the inner row, are obliterated over a broad central section of the marginal band.

Species' range

The Monarch is essentially a New World species flying in North, Central and South America, and the Caribbean, but since the mid-nineteenth century it has greatly expanded its range and is now found in the Azores, Canary Islands, Madeira, Hawaii, Indonesia, Australia and New Zealand. It is a vagrant to western Europe and has recently become established in southern Spain.

Cayman Islands distribution

Grand Cayman, Little Cayman, Cayman Brac

Habitat

Danaus plexippus is most often seen in parks and gardens where *Asclepias curassavica* (Red Top) is grown. An individual butterfly will often frequent the same area on several consecutive days.

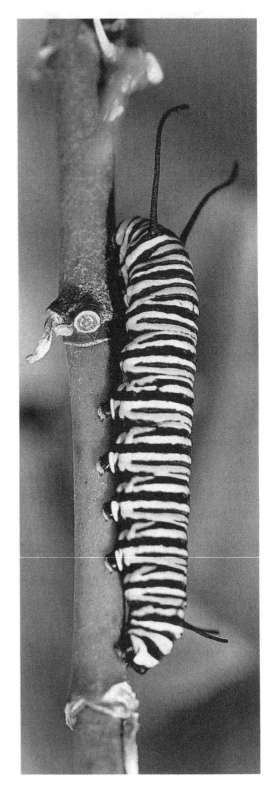

Monarch larva, CB (28.i.2008), RRA

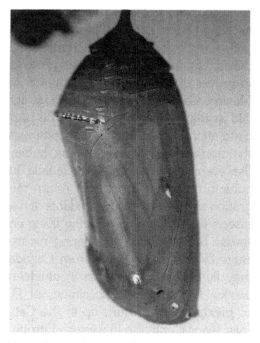

Monarch pupa, CB (30.i.2008), RRA

Larva on Red Top (Asclepias curassavica), attacked by the paper wasp Polistes major, GC (28. xii.2006), AS

History

Carpenter & Lewis (1943) record the Monarch from Grand Cayman and Cayman Brac, describing it as not scarce but very limited in its distribution. Their material appears to have belonged to *D. p. megalippe* and probably came from breeding populations on the Cayman Islands. Between 1975 and the end of the century, only a few individuals were noted on Grand Cayman, and none on Little Cayman or Cayman Brac. A Monarch collected in October 1990 was clearly the migratory North American form (Miller & Steinhauser 1992). With the advent of the new millennium there came an increase in the number of Monarchs seen in the Cayman Islands, those examined all being *D. p. megalippe*. On Grand Cayman they are now not infrequent in parks in George Town and East End, and in 2008 they were quite plentiful, although very local, on Cayman Brac, and recorded for the first time on Little Cayman.

Biology

The pale green eggs are rather more than one and a half times as tall as broad, longitudinally ridged, and laid singly on leaves of milkweeds (Apocynaceae). In Grand Cayman the caterpillars are known to feed on *Asclepias curassavica* (Red Top or Scarlet Milkweed), and on the imported African species *Calotropis gigantea* (Giant Milkweed) and *C. procera* (French Cotton) but preferring the *Asclepias*. The caterpillars are warningly coloured with each segment ringed in white, black and yellow. There are two pairs (not three as in the next two species) of fleshy, mobile, tubercular processes, the longer pair on the second segment after the head and the other pair near the posterior end on the eleventh segment. Larvae are not avoided by all predators; a large caterpillar has been seen to be killed and eaten by a paper wasp *(Polistes major)*, and three assassin bugs (Reduviidae) have been found feeding on a single dead caterpillar. The pupa is dumpy and smooth, pale green with golden spots. It is suspended, often from the food-plant, by its posterior cremaster and is unsupported by a girdle. Adult butterflies fly quite slowly with frequent gliding, advertising their warning colouration, and they are quite easily approached when feeding at flowers. Nectar may be taken from flowers of the larval food-plants, and also from *Asystasia gangetica, Bidens alba, Bougainvillea* and many others. Mating behaviour in *D. plexippus* is considerably less refined than in the following two species of *Danaus* which both employ sex pheromones sprinkled by the male, from hair pencils, over the female. The male Monarch does not seem to use its undersized hair pencils during courtship to induce receptivity in the female. Instead, the male drives the female to the ground and forcibly copulates with her. This shift in the Monarch from the sophisticated, chemically mediated courtship displayed by other Danaidae, to physical, forced copulation, is remarkable, but the reason for its evolution is unexplained. It has been suggested that it may be related to the migratory habit of North American Monarchs, males with limited fat reserves after completing a southward autumn migration and overwintering may improve their prospects of perpetuating their genes by forcing themselves on females before embarking on an exhausting northward spring migration (Boppré 1993).

The North American subspecies, *D. p.* *plexippus*, is a well known and spectacular migrant (Urquhart 1976). In spring, small groups or individuals move north from dense, communal overwintering roosts in groves of conifers and *Eucalyptus* in Mexico, California and Florida. Eggs are laid as the insects fly north and the next generation of adult butterflies continues the northward migration. Some butterflies eventually reach southern Canada. In autumn there is a spectacular return migration southward, the descendants of the insects that flew north in spring flying *en masse* back to the overwintering roosts. Butterflies migrating south from Canada may fly almost 2,000 miles in nineteen weeks. The occasional specimens of *D. p. plexippus* which turn up in the Cayman Islands, and also in western Europe, are probably migrating individuals that have gone astray. However, the majority of Monarchs seen in the Cayman Islands belong to the non-migratory subspecies *D. p. megalippe*.

Queen

Danaus gilippus (Cramer, 1775)
Plate I (3-5)

Recognition
FWL 36-41 mm. The chocolate brown wings have black margins, broadest on the hindwings, and small white spots in the margins and on the apical half of the forewing. The veins are black on the underside of the hindwing, narrowly outlined with white, and the male sex brand is larger and more conspicuous than in the Monarch. The Queen is similar to the Soldier, described next, but can be distinguished by its very slightly brighter chocolate brown ground colour, the white

Male Queen on Asclepias, *GC (12.iii.2004), MLA*

outlines to the veins on the underside of the hindwing, the narrower black margin on the forewing upper surface which does not surround the inner of the two rows of white marginal spots, and by the absence of faint, pale spots on the under surface of the hindwing. The Queen has a postdiscal row of three white spots in spaces 1, 2 and 3 on the forewing upper surface, and is on average slightly smaller than the Soldier.

Subspecies

The subspecies in the Cayman Islands is *D. g. berenice* (Cramer, 1779) which is found also in Florida, the Bahamas and Cuba. A paler subspecies, *D. g. jamaicensis* (Bates, 1864) flies in Jamaica, and *D.*

g. cleothera (Godart, 1819), which has the front half of the forewing darker than elsewhere on the upper surface, is found principally in Hispaniola.

Species' range

A butterfly found through the south-eastern United States, Central America and in South America southward as far as Argentina. Its island range is the Bahamas and Greater Antilles and associated islands, but it is absent from most of the Lesser Antilles.

Cayman Islands distribution

Grand Cayman, Little Cayman, Cayman Brac

Habitat

This is a butterfly of open places, to be seen on uncultivated land and in parks and gardens. It was stated by Carpenter & Lewis (1943) to be rare in the dry interior of Grand Cayman where specimens were often dwarfed.

History

The Queen is usually common and widespread on Grand Cayman, but sporadic in occurrence on Little Cayman. It has not been recorded from Cayman Brac since 1938 (Carpenter & Lewis 1943), but this is probably because it has been overlooked.

Biology

The yellowish green eggs are taller than those of the Monarch, but as in that species they are laid singly on leaves of Apocynaceae. *Asclepias curassavica* (Red Top), less commonly *Calotropis gigantea* (Giant Milkweed) and *C. procera* (French Cotton), are known as food-plants on Grand Cayman where *A. curassavica*

Pair of Queens in copulation, male above, GC (1.i.2008), JG

may sometimes be defoliated by the larvae. Elsewhere, *Nerium oleander* (Oleander) and species of *Metastelma (= Cynanchum)* and *Sarcostemma* are eaten. On Grand Cayman *Sarcostemma clausum* (White Twinevine) is very probably a larval food-plant, the butterflies being commonly seen on uncultivated land overgrown by the plant. The fully grown larva is whitish with broad and narrow purple-black rings, the broad rings with yellow dorsal marks. The yellow is brighter in male than in female larvae (Hopf 1954). There are three pairs of blackish, fleshy, mobile processes. These are on the second and eleventh body segments, as in the Monarch caterpillar, with an additional pair on the fifth body segment. Each of these processes has a red mark at its base. The pupa of the Queen is more slender than that of the Monarch, pale green with some anterior golden tubercles. As in all Danaidae, it hangs head downwards.

Adults emerge about ten days after pupation and can be found at all times of year. There are reports of migrations of *D. gilippus,* but these are not nearly so spectacular as those of *D. plexippus.* The Queen flies quite slowly but without the majestic gliding flight of the Monarch. Prior to mating, the male *D. gilippus* flutters above the flying female and scatters scent scales from his hair pencils onto her antennae. This has the affect of making her receptive and she alights. The male hovers over her with hair pencils still everted. Eventually the female closes her wings, whereupon the male alights beside her and copulation ensues. The pair remain coupled for an average of 7 to 8 hours (Pliske 1973). In many butterflies, a single mating is sufficient to ensure that all eggs will be fertilized, and in these species copulation renders females unreceptive to further mating attempts. Queens are relatively long-lived butterflies, however, and a female may be inseminated by several males during her lifetime.

On Grand Cayman, nectar plants of *D. gilippus* include the larval food-plant *Asclepias curassavica*, and *Ageratum littorale, A. conyzoides, Bidens alba, Borrichia arborescens, Gynura aurantiaca, Jatropha integerrima, Lantana camara* and *Stachytarpheta jamaicensis.*

Queen larva, GC (25.i.2006), MLA

Soldier

Danaus eresimus (Cramer, 1777)

Plate I (6-8)

Recognition

FWL 38-42 mm. The Soldier is very like the Queen (see above) but may be identified by having on the forewing two rows of white dots lying within the dark margin and larger but less distinct pale postdiscal spots in spaces (1), 2 and 3 in the female (these spots usually absent in the male). There is no white lining to the black veins on the under surface of the hindwing and, most noticeably, there is an arc of four or five oval, pale spots between the veins on the outer half of the hindwing. These latter spots may be faint, but are present on both surfaces of the wing, and are the most reliable character for distinguishing the Soldier from the Queen. Lewis in Carpenter & Lewis (1943) noted that the Soldier 'was always easily distinguished on the wing from *berenice* [Queen], the latter being much darker'.

Subspecies

Danaus eresimus tethys Forbes, 1943 occurs in the Cayman Islands, Cuba, Jamaica, Hispaniola, Grenada and quite recently has colonized the southern United States. The nominate subspecies is a South American insect that occasionally turns up in the southern West Indies. Ackery & Vane-Wright (1984: 110) give both *D. e. tethys* and *D. e. montezuma* Talbot, 1943 as occurring on Grand Cayman (Lamas, unpublished), but the coexistence for any length of time of two subspecies on a small island is unlikely.

Species' range

The species ranges from the southern United States, through Central America and the West Indies, and into South America as far as the Amazon Basin.

Cayman Islands distribution

Grand Cayman, Little Cayman

Habitat

The Soldier occurs in similar habitats to the Queen (see above) but may be more tolerant of very dry places such as beach ridges. Like the Queen, the Soldier is often to be seen on uncultivated land that has become overgrown by *Sarcostemma clausum*.

History

A common and widespread butterfly on Grand Cayman since first recorded there in 1938 (Carpenter & Lewis 1943), and generally rather more abundant than *D. gilippus*.

It is present on Little Cayman (F. Roulstone, pers. comm.) but has not been recorded from Cayman Brac.

*Female Soldier on Red Top (*Asclepias curassavica*), GC (1.ii.2006), MLA*

Biology

The egg is orange and resembles the Milkweed Aphid *(Aphis nerii)* which is a pest of *Asclepias curassavica* (Red Top), the larval food-plant of *D. eresimus*. The fully grown larva is greenish yellow with yellow, white and dark maroon rings, and three pairs of fleshy, motile processes. These are positioned as in the larva of the Queen, but they have maroon rather than red bases, and the posterior pair is slightly shorter than the others. The caterpillar of the Queen has the anterior pair of processes distinctly longer than the pair on the fifth body segment, and these in turn are a little longer than the pair on body segment eleven (Smith *et al.* 1994). On Grand Cayman larvae have been found feeding on *Asclepias curassavica*. Elsewhere other Apocynaceae such as *Metastelma (= Cynanchum)* and *Sarcostemma clausum* (White Twinevine) are used. The pupa is similar to that of *D. gilippus*.

Butterflies take nectar from a similar range of plants to the Queen, and it is not uncommon to see the two species feeding together. On Grand Cayman nectaring has been observed on *Ageratum littorale*, *A. conyzoides*, *Asclepias curassavica*, *Bauhinia divaricata*, *Bidens alba*, *Coccoloba uvifera*, *Colubrina cubensis*, *Duranta erecta*, *Gynura aurantiaca* and *Jatropha integerrima*. The flight of *D. eresimus* is said to be faster and slightly higher than that of *D. plexippus* or *D. gilippus* (DeVries 1987), but unlike the two preceding species of *Danaus*, *D. eresimus* is not known to migrate. Biting midges *(Forcipomyia)* of the family Ceratopogonidae have been seen to draw haemolymph from the wing veins of the Soldier (Lane in Vane Wright & Ackery 1984).

Male Soldier nectaring at White Button (Spilanthes urens), *GC (20.x.2007), KDG*

Brush-footed Butterflies

Nymphalidae

The reduction of the front pair of legs to brush-like appendages, which are not used in walking, is a character of adult Danaidae and Nymphalidae. Opinion is divided, but some lepidopterists consider these, together with Ithomiidae and Satyridae (see pages 131, 132) to be no more than subfamilies of Nymphalidae. Another feature which distinguishes these four families from most other butterflies is the absence of a median silken girdle supporting the pupa. The pupa, in consequence, hangs head down, attached to the substrate only at the posterior end where the cremaster is fastened to a silken pad.

Nymphalidae, even in the restricted sense used here, is a very large family. It is found world-wide but probably reaches its greatest diversity in the Neotropical Region. It is well represented in the West Indies where, in number of species, it is second only to Hesperiidae (Skippers). Most nymphalids are medium-sized, strong-flying, colourful butterflies. The sexes are usually similar in wing pattern, but reduction of the front legs is more extreme in males.

Eggs of Nymphalidae are usually barrel-shaped and ribbed, with intricate surface sculpture, although sometimes, as in *Memphis*, they are spherical and smooth. All nymphalid larvae feed on dicotyledons, unlike the grass-feeding larvae of Satyridae, and they are nearly always clothed with black, branched spines and sometimes other processes and filaments as well. The pupa bears two short horns on the head and it is often decorated with shining golden spots (giving it the alternative name of chrysalis, meaning golden).

Cuban Red Leaf Butterfly

Anaea troglodyta (Fabricius, 1775)
Plate I (9-11)

Recognition
FWL 32-42 mm, the female distinctly larger than the male. The forewing is falcate, its costal margin slightly convex and apex acutely pointed, and the hindwing is tailed. The upper side is mainly red with brown markings. In the female the ground colour is more orange-red, and the brown wing margin is broader than in the male, broadest at the forewing apex and gradually narrowing posteriorly, extending onto the hindwing but disappearing just before

Male Cuban Red Leaf Butterfly, GC (8.ii.2006), MLA

Female Cuban Red Leaf Butterfly, GC (8.ii.2006), MLA

Subspecies
A. t. cubana (Druce, 1905) is known from Cuba, Isle of Pines and Grand Cayman. Other subspecies include *A. t. portia* (Fabricius) from Jamaica, *A. t. borinquenalis* Johnson & Comstock from Puerto Rico and the nominate *A. t. troglodyta* from Hispaniola (Scott 1986a). All are sometimes regarded as full species following the discovery of small differences in the male genitalia (Comstock 1961).

Species' range
Anaea troglodyta occurs in the Greater Antilles and northern Lesser Antilles.

Cayman Islands distribution
Grand Cayman only, although the larval food-plant is common on all three Cayman Islands and the Cuban Red Leaf Butterfly could be a future colonist of the Sister Islands.

the tail. Internal to the marginal band in the female is a series of brownish yellow spots which become progressively larger from the anterior one-third of the forewing to the inner margin of the hindwing. These spots are present in the male but faint, and on the hindwing only. The under side is mottled greyish brown with numerous dark flecks, and the forewing is flushed with orange-red.

Larva on leaf-fold of Rosemary (Croton linearis), GC (30.i.2006), RRA

Larva showing head features, GC (9.ii.2006), RRA

37

Habitat

This is a butterfly usually found near trees, along woodland edges or tall hedges, in open woodland and in parks and gardens with mature trees. It is seldom seen far from the larval food-plant, *Croton linearis* (Rosemary).

History

Anaea troglodyta was first discovered in the Cayman Islands by Dr E. J.Gerberg on 20 September 1983 at Boatswain Point, Grand Cayman; about a dozen specimens were seen and were thought to be breeding in the area (Gerberg, personal communication 1987). In August 1985, Askew found several fresh specimens at Botabano and Mount Pleasant, in the north-west of Grand Cayman, and being unaware of Gerberg's initial discovery, reported the species as new to the Cayman Islands (Askew 1988). The species was clearly breeding in 1985, and in November of that year it was found independently by Albert Schwartz and Fernando Gonzalez in George Town and at Boatswain Bay (Schwartz *et al.* 1987). Since 1985, *A. troglodyta* has spread throughout the western peninsula of Grand Cayman, and later moved eastwards. Discontinuous observation at the end of the century showed it to have colonized parks and gardens in George Town and South Sound by 1995, reaching Midland Acres by 1997. In 2002 it was found in plenty at the southern end of the Mastic Trail and in 2004 or earlier it had reached the northern end of the Mastic Trail and was in Queen Elizabeth II Botanic Park. In 2006 specimens were seen east of Old Man Bay on the north coast, at Cottage in the south and at High Rock and Great Beach in the eastern interior. The Cuban Red Leaf Butterfly is now common throughout Grand Cayman, and it clearly outnumbers the much longer established Cuban Brown Leaf Butterfly (see below).

Male Cuban Red Leaf Butterfly on perch, GC (10.xi.2007), KDG

Biology

The larva of *Anaea troglodyta* feeds on *Croton linearis* (Rosemary) (Euphorbiaceae), a common shrub in light woodland, parks and gardens. Eggs are laid singly on the underside of a *C. linearis* leaf and the larvae live solitarily in tubular shelters or hammocks constructed from leaves of the food-plant spun together longitudinally. These are quite easily found, and when inhabited the head of the larva can be seen blocking the entrance. At night, the larvae leave their shelters to feed. Entire leaves are consumed so that no partly eaten leaves can disclose the whereabouts of the larvae. A fully grown caterpillar is grey-green with a brown m-shaped mark dorsally on the first abdominal segment. Head and thorax are covered in very small, white papillae. The greenish head capsule is crowned with large white papillae and the face has three pale chevrons. There are three pairs of red-

Pupa of Cuban Red Leaf Butterfly with adult wing pattern visible through wing-case, GC (7.ii.2006), RRA

brown stemmata (simple eyes) and above them a dorsal pair of dark brown protuberances. The whole bears a fanciful resemblance to the head of a walrus. The pupa is light green and smooth, usually suspended from a leaf of *Croton* by a black cremaster attached to a pad of silk. A large pupa, measuring 19 mm x 11 mm, produced a female *A. troglodyta* after eleven days. This butterfly emerged at 11.00 h and took its maiden flight at 16.00 h.

The adult butterfly has been observed in every month except May and June. It makes rather short flights and passes long periods perched with wings closed on the top of horizontally growing twigs or branches, or sometimes on logs and stones on the ground. When on a branch it much resembles a dead leaf, and on the ground its hindwing under surface is very cryptic. Upon alighting the butterfly usually briefly opens its wings once. Perching butterflies can often be closely approached, sometimes even touched, relying on their stillness and concealing colouration to avoid detection. In flight, however, the red upper surface flashes conspicuously. Butterflies have not been seen to visit flowers, but they will feed at fermenting fruit.

Cuban Brown Leaf Butterfly

Memphis verticordia (Hübner, 1831)
Plate I (12-14)

Recognition
FWL 23-27 mm (Little Cayman), 29-34 mm (Grand Cayman); insects from Grand Cayman are larger than those from the Sister Islands, a situation similar to that of *Heraclides andraemon* (page 110). The forewing upper surface is dark chestnut-brown basally and broadly brown on the outer margin, these two areas separated by a dark brown zone. The hindwing is tailed, similar in colour to the forewing on the upper surface, but with a broader basal chestnut-brown area and two small black spots with white pupils near the base of the tail. The under surface is dull brown with an intricate pattern of fine black lines, and two eye-spots between the base of the tail and the tornus, as on the upper surface. These eye-spots and the tail lie in a quite sharply defined buff coloured area. The sexes are alike except that the female may have a pair of small, yellowish, subapical spots on the forewing upper surface close to the costal margin.

Subspecies
Memphis verticordia danielana (Witt, 1972) is a Caymanian endemic and may be given the vernacular name of Cayman Brown Leaf Butterfly. It is not, however, very strongly differentiated from two other subspecies, *M. v. echemus* (Doubleday, 1849) in Cuba and the Isle of Pines and *M. v. bahamae* (Witt, 1972) in the Bahamas. *M. v. danielana* was described from specimens collected by the Oxford University expedition and now in the Natural History Museum, London. The holotype is a male collected on 29 June 1938 at the west end of George Town,

Grand Cayman, but specimens from the Sister Islands are also included in the type series. *M. v. danielana* is named by Witt (1972) for Franz Daniel, his entomology teacher in Munich.

Species' range
The Bahamas, Cuba, Isle of Pines and Cayman Islands

Cayman Islands distribution
Grand Cayman, Little Cayman, Cayman Brac

Habitat
Memphis verticordia is usually seen in open woodland and at the edges of woods, in similar habitats to those of *Anaea troglodyta* except that it is less frequent in managed parks and gardens. It is a butterfly very much associated with trees and bushy places.

Male Cayman Brown Leaf Butterfly on perch, GC (9.ii.2006), RRA

History
This species has apparently been a continuous resident on all three Cayman Islands since first recorded in 1938, although it is never a common butterfly and is usually seen singly or in very small numbers.

Biology
Memphis verticordia has been observed in the Cayman Islands laying its smooth, spherical eggs on *Croton nitens* (Wild Cinnamon), and in Cuba caterpillars have been found on *C. lucidus* (Fire Bush). *C. linearis* (Rosemary) does not seem to attract the butterfly. Adult butterflies can probably be found all year round. They start flying early in the day, soon after sunrise, and continue until dusk, but during the heat of the day they retreat into the shade of bushes, and flight activity is greatly reduced from about 11.00 h until 15.00 h (Carpenter & Lewis 1943, Askew 1994). Much time is passed perching on twigs and branches with wings closed, often deep in thickets, where the butterflies are exceptionally well concealed by the cryptic pattern of their hindwing under surfaces. In flight, *Memphis verticordia* is less conspicuous than *Anaea troglodyta* because of its darker upper surface, but if it is seen to alight by a bird or *Anolis* lizard, any attack is likely to be deflected to the eye-spots and tails on the hindwings. A large proportion of butterflies have damage to these areas (Carpenter & Lewis 1943) and nine out of sixteen butterflies examined on Little Cayman in 1975 had one or both tails missing.

Memphis verticordia seems seldom to take nectar (*Petitia domingensis* has been noted as a nectar source), but it will feed at exuding tree sap and fallen, fermenting fruit. On Grand Cayman it has been seen on overripe mangoes, breadfruit,

Eggs of Cayman Brown Leaf Butterfly on leaf of Wild Cinnamon (Croton nitens), GC (10.v.2007), AS

Surinam cherries and grapefruit on the ground. The butterflies are attracted to alcohol, and the only occasion on which several have been seen together was on Little Cayman in 1975 when seven or eight were fluttering around the collecting head of a Malaise trap charged with ethanol.

Antillean Ruddy Daggerwing

Marpesia eleuchea Hübner, 1818
Plate II (1)

Recognition
FWL 33-36 mm. The hindwing of *Marpesia* has a long tail, considerably longer than in *Anaea* or *Memphis* (see above), and the tornal angle is also produced. The upper surface of *Marpesia eleuchea* is tawny orange with narrow, black, transverse lines. The underside is various shades of purplish brown and lilac, and resembles a dead leaf.

Subspecies
Caymanian specimens of *M. eleuchea*

are of the nominate subspecies which is found in Cuba and the Isle of Pines. Elsewhere, three other subspecies are recognized: *M. e. dospassosi* Munroe, 1971 in Hispaniola, *M. e. pellenis* (Godart, 1824) in Jamaica and *M. e. bahamensis* Munroe, 1971 in the Bahamas.

Marpesia petreus (Cramer, 1776) is a closely similar species occurring in Texas, Florida, Central and South America, Puerto Rico, Mona and the Lesser Antilles. It differs from *M. eleuchea* in having a longer tail, more produced tornal angles and forewing apex, and straight transverse black lines on the forewing. In *M. eleuchea* the only complete transverse black line is angled basally over its anterior one-third. Some authorities treat *M. eleuchea* as a subspecies of *M. petreus*.

Species' range
Marpesia eleuchea occurs in the Bahamas, Cuba, Isle of Pines, Hispaniola and Jamaica, and is recorded as a vagrant in southern Florida and Grand Cayman.

Cayman Islands distribution
Grand Cayman (vagrant and temporary resident)

Habitat
This is a woodland butterfly which was seen on Grand Cayman in gardens with mature trees.

History
Marpesia eleuchea was discovered on Grand Cayman by Ann Stafford on 20 December 2001 when it was seen and photographed on wooded land adjacent to her garden in George Town. At least two individuals were present up to 27 December. Early in 2002, between 3 January and 16 February, there were sporadic

Antillean Ruddy Daggerwing, GC (3.i.2002), AS

Male feeding at blossom of Cajon (Colubrina cubensis), GC (3.i.2002), AS

sightings of butterflies up to a half mile away from the original site, and then on 9 March 2002 two specimens reappeared at the place where the species had first been found. These were noticeably smaller than the original specimens and probably their offspring. Sightings continued up to 26 April, but *Marpesia eleuchea*, to our knowledge, has not since been seen on Grand Cayman.

Biology

The larval food-plants of *Marpesia eleuchea* in Cuba are the figs *Ficus aurea* and *F. carica*. Hernández (2004) describes the caterpillar as having a milky white head ornamented by two long, black horns with globular tips, and the sides of the body marked with violet stripes. The pupa is green.

Butterflies in Grand Cayman were regularly observed taking nectar from *Colubrina cubensis,* and they also frequented a Fiddlewood Tree (*Petitia domingensis*).

They have a slow, gliding flight and, unlike the two Leaf Butterflies described above, they often perch with wings spread. One of the Grand Cayman insects was without a large part of its hindwings, including both tails; it had presumably survived a lizard attack.

Many-banded Daggerwing

Marpesia chiron (Fabricius, 1775)
Plate II (2)

Recognition

FWL 28-30 mm. *Marpesia chiron* is another butterfly with long hindwing tails, but it is smaller than *M. eleuchea* (see above) and readily distinguished from it by its four complete and broad dark brown bands on the upper surface. These dark bands are fully as broad as the pale bands between them, and they are continuous from the forewing onto the hindwing.

There are three small, white spots in the submarginal dark band near the apex of the forewing. On the underside, both front and hindwings are whitish in their basal halves, brownish distally, and with finer dark, longitudinal stripes.

Subspecies
No subspecies have been described. The lack of divergence of West Indian populations of *M. chiron* from those on the mainland is perhaps because of frequent invasion of the islands from Central America (Smith *et al.* 1994).

Species' range
This butterfly is distributed through Central and South America to Argentina, regularly migrating north to southern Texas and Florida, occasionally as far as Kansas. In the West Indies it is common in Cuba and the Isle of Pines, known from a few localities on Hispaniola (Dominican Republic), and very scarce in Jamaica and Puerto Rico (Smith *et al.* 1994).

Cayman Islands distribution
Grand Cayman (vagrant)

Habitat
Marpesia chiron in Cuba is found in woodland clearings and edges.

History
On 5 December 2005, a specimen of *M. chiron* was photographed in south George Town by Ann Stafford. The insect was missing the tails on both hindwings. A year later, on 8 December 2006, another was photographed in almost the same spot. This second specimen was without the tail on its left hindwing. These are the only records to date of the species in the Cayman Islands.

Many-banded Daggerwing with one tail missing, the second specimen seen on GC (8.xii.2006), AS

Biology
The larva of *M. chiron* feeds on various Moraceae (*Maclura* (=*Chlorophora*), *Ficus*, *Morus*, *Artocarpus*). It is spiny, yellow-orange (greenish on the sides) with red, transverse marks and two black, longitudinal streaks on the back (Brown & Heineman 1972).

The adult butterfly feeds on fallen fruit, faeces and at damp earth, as well as sometimes nectaring at flowers. The specimens found on Grand Cayman appeared to be attracted to *Casearia aculeata* (Thorn Prickle) (Salicaceae). In Costa Rica, DeVries (1987) describes males visiting 'wet sand by the hundreds [resembling] a great gray-violet cloud as they swirl around each other'. On the mainland, *M. chiron* is known to undertake mass migrations, often in company with the large moth *Urania fulgens* (page 158), and it is capable of adjusting its flight course to compensate for wind drift (Srygley *et al.* 1996). *M. chiron* may have just one generation in a year.

Haitian Cracker, Click Butterfly

Hamadryas amphichloe (Boisduval, 1870)
Plate II (3)

Recognition
FWL 36-40 mm. The upper surface has a complex pattern of grey and white markings and wavy black lines. The forewing has large white spots on its outer two-thirds, and there is a series of inconspicuous submarginal eye-spots on both fore- and hindwing. These eye-spots are merely black circles with white centres, the black broadest on the inner sides and edged internally by a narrow brown crescent on the two hindmost eye-spots of the hindwing. The under surface is predominantly whitish on the basal one-third of the forewing and two-thirds of the hindwing and, distal to these areas, is marked with brown surrounding irregular white spots. There are series of submarginal eye-spots, as on the upper surface.

Subspecies
Antillean populations are assigned to *Hamadryas amphichloe diasia* (Fruhstorfer,

1916). Riley (1975) refers to *diasia* as a subspecies of *H. februa* (Hübner, 1823), but this is a South American species.

Species' range
Hamadryas amphichloe has a range extending from Mexico to Peru and including the Greater Antilles (Cuba, Hispaniola, Puerto Rico, Mona Island and Jamaica). Its Antillean distribution may have expanded from Hispaniola, becoming established as a breeding species in eastern Cuba, probably in the 1930s, thence spreading westwards. It may not have invaded Jamaica before the 1980s. It occurs as a vagrant in Florida and the Cayman Islands.

Cayman Islands distribution
Grand Cayman (vagrant)

Habitat
This species is tolerant of a range of ecological conditions (Schwartz *et al.* 1989). It is usually a woodland butterfly but will fly in open situations.

Haitian Cracker resting head-down on tree trunk, GC (13.iii.2005), FR

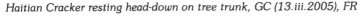

History

Like the two preceding species of *Marpesia*, *H. amphichloe* is known in the Cayman Islands only as a vagrant. It was not seen before 2005, but in that year there were a number of sightings of it in Grand Cayman. The first, a rather worn specimen, was photographed on 13 March 2005 by Frank Roulstone in a wooded area at Arlington Estates near Old Man Bay. It appeared among thousands of *Ascia monuste*. Another specimen was attracted to rotten bananas in the garden of Carla Reid, midway between Old Man Bay and Colliers, and Carla Reid found and photographed an apparently freshly emerged insect on 28 November 2005 at East End.

Biology

The only known larval food-plant of *H. amphichloe* is *Dalechampia scandens* (Euphorbiaceae), for example in Hispaniola (Wetherbee 1987). Other species of *Dalechampia* are used as food-plants by some other *Hamadryas* species. The immature stages are described by Wetherbee (1987).

The adult butterfly has been recorded nectaring at *Hibiscus*, *Lantana* and other flowers (Smith *et al.* 1994), but is more often seen feeding on fermenting, fallen fruit. Adults frequently rest on tree trunks, head downwards and wings spread and pressed against the bark. The clicking noise which butterflies on the mainland make in flight, and which accounts for the species' vernacular name, is rarely heard in West Indian insects. It may be an anti-predator device, but this remains to be demonstrated, as does the mechanism of click production.

Mimic

Hypolimnas misippus (Linnaeus, 1764)
Plate II (4)

Recognition

FWL 34-38 mm. The sexes are remarkably different. The male upper side is black with a large white spot in the centre of each wing, and a smaller spot near the apex of the forewing. A broad purplish zone surrounds each of the white spots. The under surface ground colour is brownish to reddish, with white spots as on the upper surface except that those in the centres of the wings are extended to reach the anterior wing margins. The outer margins of both wings on the butterfly's under surface are whitish with three narrow black lines that parallel the scalloped edges of the wings. In striking contrast to the male, the female *Hypolimnas misippus* mimics the Old World butterfly *Danaus chrysippus* (Linnaeus, 1758). The upper surface ground colour is tawny, and there is an oblique white bar marking off the darkened forewing apex. The under surface of the female is patterned much as in the male.

Subspecies

No subspecies are described.

Species' range

Hypolimnas misippus is a butterfly of the Old World tropics and subtropics that has become established in north-eastern South America (Guyana, Venezuela). It spreads north sporadically into Central America, the West Indies and occasionally reaches the United States, from whence it has been found as far north as New York. There are records of *H. misippus* from most of the Greater and Lesser Antilles (Smith *et al.* 1994). How the Mimic reached the New World is uncertain. It may have been introduced from West Africa by the slave trade, but it is a powerful flier and wind-assisted passage across the Atlantic has also been suggested.

Cayman Islands distribution
Grand Cayman (vagrant)

Habitat
In the Old World the Mimic is perhaps commonest on weedy, partially cultivated ground where its numerous larval food-plants, particularly Convolvulaceae and Portulacaceae, abound.

History
A male collected at Boatswain Point, Grand Cayman on 30 December 1975 by Dr E. J. Gerberg appears to be the only record of *Hypolimnas misippus* in the Cayman Islands (Askew 1988).

Biology
The partly gregarious larvae of *Hypolimnas misippus* feed mainly on Convolvulaceae (*Ipomoea*) and Portulacaceae (*Portulaca*), but also on Acanthaceae and Malvaceae, in Cuba. In the Old World caterpillars are known to feed on a broad range of plants belonging to several families (Vane Wright *et al.* 1977). The fully grown caterpillar is black with grey bands and branched, white spines, with two projections on its head.

The Mimic is a well-studied insect, famous for its mimicry of *Danaus chrysippus* in the Old World. It is an edible butterfly, females of which closely resemble the distasteful model. The female of *Hypolimnas misippus* is a Batesian mimic of *D. chrysippus*. Batesian mimics should be encountered by predators less

frequently than their models, otherwise the predators will not learn to associate their appearance with distastefulness. By restricting mimicry to females only, the Mimic can, in theory, approximately double the size of its population in regions where it is protected by the presence of *D. chrysippus*. Population size may be further increased by the females being polymorphic, each form imitating a different model. This occurs in West Africa where *D. chrysippus* itself is polymorphic, and nearly every form has a female form of *H. misippus* that mimics it (Owen 1971, Smith 1976). These considerations, however, are not relevant to the New World where *H. misippus* has no danaid model and females are monomorphic.

Mangrove Buckeye

Junonia evarete (Cramer, 1779)
Plate II (5,6)

Recognition
FWL 24-29 mm. The sexes are similar in *Junonia*, as they are in all the following Nymphalidae. The upper side is brown with two postdiscal eye-spots on each wing, the forewing with a very small subapical spot and a large eye-spot in the posterior half, the hindwing with the eye-spots intermediate in size between the two forewing spots, the anterior larger than the posterior and both lying just inside a pale orange-brown submarginal band. On the forewing a similarly coloured submarginal band forms a ring around the large eye-spot, and a very pale brownish fascia runs obliquely from it to the costal wing margin. This fascia comprises three transverse spots, the most posterior of which is less than twice as broad as long. Two reddish orange marks edged with

Male Mangrove Buckeye, GC (29.xii.2005), JG

Male Mangrove Buckeye with unusually pale forewing fascia, GC (22.xi.2004), AS

black lie in the forewing cell. The under side is lighter brown with the eye-spots much diminished and usually only the large forewing eye-spot conspicuous. The hindwing under side has a much reduced pale submarginal band.

Morphological characters which distinguish *Junonia evarete* from the very similar *J. genoveva* are presented below under the latter species. Our interpretation of the names *J. evarete* and *J. genoveva* follows that of Turner & Parnell (1985) and Smith *et al.* (1994). Some other authorities (e.g. Beccaloni *et al.* 2008) use the name *J. genoveva* for the Mangrove Buckeye.

Subspecies
None

Species' range
This species has a rather restricted range from southern Florida down the eastern seaboard of Central America to Honduras, and including the Bahamas, Greater Antilles, Virgin Islands and Leeward Islands to Barbuda (Smith *et al.* 1994).

Cayman Islands distribution
Junonia evarete is known from Grand Cayman and Little Cayman. It has not been recorded from Cayman Brac, which has only limited stands of the larval food-plant, but it would be surprising if it did not occur there.

Habitat
Females of *Junonia evarete* are most frequently seen in the vicinity of Black Mangrove (*Avicennia germinans*), the larval food-plant, and both males and females are particularly abundant along the dyke roads of Grand Cayman. However both sexes, but especially males, can be found on uncultivated land, beach ridges and gardens almost anywhere. They frequently bask in the sun, wings spread, on bare ground.

We quote at length from Turner & Parnell's (1985) detailed study of *Junonia* in Jamaica where they found that males 'of *J. evarete* are common along the edges of mangroves and appear to be more strongly territorial than those of *J. genoveva*. Females of *J. evarete* remain in the mangrove woodland for the most part but appear at the woodland's edge to feed or, in late afternoon, to rest on the ground in sun-lit locations. Flight in both males and females of *J. evarete* consists of a long series of powerful wing-beats as the insect rises over shrubs and trees followed by weak gliding and fluttering as the insect descends to the ground again. Males exhibit "scudding and planing" flight [as seen in *J. genoveva*] over short distances but for both sexes flight is predominantly "soaring and fluttering".' Smith *et al.* (1994) found similar flight behaviour of *J. evarete* in Florida.

History

Carpenter & Lewis (1943) recorded *Junonia evarete*, under the name *Precis lavinia f. genoveva*, from Grand Cayman and Little Cayman and, at the same time, *J. genoveva* (as *Precis lavinia f. zonalis*) from Grand Cayman and Cayman Brac. Subsequently, both species have been reported from Grand Cayman with *J. evarete* usually the more numerous. Only *J. genoveva* has been noted on Cayman

Mangrove Buckeye nectaring at Vervine (Stachytarpheta jamaicensis), GC (29.i.2006), RRA

Brac. On Little Cayman in 1975 *J. genoveva* was abundant but *J. evarete* was not seen (Askew 1980), the reverse of the situation found in 1938 (Carpenter & Lewis 1943). It may well be that both species of *Junonia* are resident on all three Cayman Islands, but the presence of *J. evarete* on Cayman Brac and its persistence on Little Cayman require confirmation.

Biology

Females of *J. evarete* have been observed on Grand Cayman ovipositing on the upward pointing 'breathing' roots or pneumatophores of *Avicennia germinans* (Black Mangrove), the only known larval food-plant. Turner & Parnell (1985) note that the larvae of *J. evarete* are larger at all stages than those of *J. genoveva* and the 'bases of the mid-dorsal scoli [knobbly, fleshy, finger-like processes bearing long setae] of living *J. evarete* larvae are iridescent turquoise whereas those of *J. genoveva* larvae are iridescent purple. Larvae of both species have eight rows of scoli on the thoracic segments and nine longitudinal rows of scoli on the abdominal segments, including a mid-dorsal row that is absent on the thorax.' The caterpillar is predominantly black and the head bears a pair of long scoli which are slightly swollen at their tips. The pupae of *J. evarete* are described as being consistently darker than those of *J. genoveva* with black markings, whereas those of *J. genoveva* have pink, white and green markings.

Adult butterflies have been seen at all times of year. Nectar plants recorded on Grand Cayman include *Bidens alba* and *Stachytarpheta jamaicensis*. Males are territorial, often resting on bare ground in full sun.

Caribbean Buckeye, Tropical Buckeye

Junonia genoveva (Cramer, 1780)
Plate II (7-9)

Recognition

FWL 24-29 mm. *Junonia genoveva* is very similar to *J. evarete* (above), and there is some confusion in the literature over the names applied to the two species. Differentiating between the two Buckeye species is likely to be the most difficult identification problem posed by Cayman butterflies. Characteristically marked specimens are fairly straightforward, but there is considerable intraspecific variation which makes butterflies more or less intermediate in appearance between typical examples of the two species difficult to name. The problem is compounded by the apparent existence of seasonal forms of both species. Turner & Parnell (1985) made a very thorough study of *J. genoveva* and *J. evarete* in Jamaica, and their interpretation of the two species, which hitherto had not been satisfactorily distinguished, is followed here. Features separating adults of *J. genoveva* and *J. evarete* in the Cayman Islands are tabulated below; these are based on Turner & Parnell's (1985) table of differences in Jamaican populations. An average difference between the two species that is not listed is the number of spines or teeth on the ends of the valves in the male genitalia, the mean valve spine number for *J. evarete* in Jamaica being 17.4 (range 12 to 21 in 15 specimens) and for *J. genoveva* 27.6 (range 18 to 35 in 15 specimens).

Male Caribbean Buckeye, GC (21.i.2008), RRA

Pale forewing fascia

J. genoveva Whitish, relatively broad and not much constricted anterior to the large eye-spot, the third inter-vein spot from the costal margin more than twice as broad as long

J. evarete Tawny, rarely partly white, narrower and hardly extending into the orange ring surrounding the large eye-spot, the third spot from the costal margin less than twice as broad as long

Antennal club colour

J. genoveva Whitish, some brown scales only in a narrow patch on dorsal surface

J. evarete Mainly brownish, white only immediately behind club apex and in a narrow band on the ventral surface; the stem of the antenna is also darker than in *J. genoveva*

Submarginal, orange-brown band of hindwing

J. genoveva Reduced in male, occasionally also in female

J. evarete Relatively broad in both sexes

Relative sizes of the hindwing eye-spots differ only slightly and not invariably between the two species in the Cayman Islands, the anterior spot tending to be larger in *J. genoveva* than in *J. evarete*.

J. coenia Hübner, 1822, a possible vagrant to the Cayman Islands (page 132), resembles *J. genoveva* in having a broad,

Larva of Caribbean Buckeye, CB (28.i.2008), RRA

white forewing fascia, but it may be distinguished by the very large anterior hindwing eye-spot which exceeds the diameter of the forewing eye-spot and is more than twice the diameter of the posterior hindwing eye-spot. The white forewing fascia in *J. coenia* extends down the inner margin as well as the outer margin of the posterior ocellus, and surrounds it but for a short posterior segment of its periphery.

Subspecies
None

Species' range
Central America, northern parts of South America, Florida (vagrant), Bahamas, Greater and Lesser Antilles

Cayman Islands distribution
Grand Cayman, Little Cayman, Cayman Brac

Habitat

Junonia genoveva is not associated with mangroves and it usually flies in drier places than *J. evarete,* but both species frequently occur together. Only once on Grand Cayman, in 2002 at High Rock, have we found *J. genoveva* to be the more numerous of the two species. On Little Cayman, *J. genoveva* displays a distinct preference for beach ridges. Carpenter & Lewis (1943) present a different picture of habitat preferences shown by these two species of *Junonia*, writing that *J. genoveva* (as *Precis lavinia* f. *zonalis*) was found in meadow and pasture whilst *J. evarete* (as *P. lavinia* f. *genoveva*) occurred along the tops of the beaches and within one hundred yards of the shore.

History

Junonia genoveva was first recorded from Grand Cayman and Cayman Brac by the 1938 Oxford University expedition (Carpenter & Lewis 1943, under *P. lavinia* f. *zonalis*). In 1975 it was observed on Cayman Brac and Little Cayman, the only *Junonia* then seen on the Sister Islands, but not on Grand Cayman where only *J. evarete* was found (Askew 1980). Subsequently both species have been reported from Grand Cayman but only *J. genoveva* from Cayman Brac (Schwartz *et al.* 1987, Miller & Steinhauser 1992).

Biology

Caterpillars of *J. genoveva* have been found feeding on *Stachytarpheta jamaicensis* (Vervine or Porterweed) on Grand Cayman and Little Cayman, and other recorded food-plants are *Blechum pyramidatum* (= *brownei*) (Blechum), *Ruellia tuberosa* (Duppy Gun or Heart Bush) and *Stemodia maritima* (Seaside Twintip). They will not eat Black Mangrove, and the difference in larval food-plants between *J. genoveva* and *J. evarete* is possibly fundamental to their coexistence.

Adult butterflies may be found at any time of year. On Grand Cayman they have been noted feeding at *Bidens alba*, *Duranta erecta* and *Stachytarpheta jamaicensis*. Lewis in Carpenter & Lewis (1943) claims that he found it easy to distinguish *J. genoveva* and *J. evarete* by their flight, *J. genoveva* flying 'erratically and for short distances only' and *J. evarete* taking 'long, rapid and fairly straight flights'. These observations correspond with Turner & Parnell's (1985) account of both sexes of *J. genoveva* flying close to the ground with 'a short series of powerful wing-beats, followed by a longer series of wing-beats of small amplitude resulting in a "scudding and planing" flight', whilst in *J. evarete* 'flight is predominantly "soaring and fluttering"' as described above (page 48).

White Peacock

Anartia jatrophae (Linnaeus, 1763)
Plate II (10,11)

Recognition

FWL 26-32 mm. *Anartia jatrophae* is a whitish butterfly with a complex pattern in grey to light brown and three pairs of small, black pseudo-eyespots, one on the forewings and two on the hindwings. The hindwing margin is scalloped with a short, stubby tail. Females are somewhat larger than males, and slightly darker, but otherwise the sexes are similar.

Subspecies

Anartia jatrophae jamaicensis Möschler, 1886 is the subspecies recorded from

White Peacock, GC (2.ii.2006), MLA

Jamaica, Grand Cayman and possibly Cayman Brac (Smith *et al.* 1994). There is, however, considerable overlap in appearance between the several described West Indian subspecies of *Anartia jatrophae*, and some seasonal variation complicates the situation, so that subspecific division of the species seems of limited value.

Species' range
The White Peacock is distributed from the southern United States, the Bahamas and West Indies, through Central America and south to Argentina.

Cayman Islands distribution
Grand Cayman, Cayman Brac

Habitat
This is a butterfly of open spaces with bare ground and weedy, secondary vegetation. It is also common in pastureland, in parks and gardens, and on the landward aspect of beach ridges.

History
In 1938 the Oxford University Biological expedition found *Anartia jatrophae* plentiful on Grand Cayman, but only a solitary, battered individual was seen on Cayman Brac, and none was observed on Little Cayman (Carpenter & Lewis 1943). It is still a common butterfly on Grand Cayman. On a single day spent on Cayman Brac by the Royal Society – Cayman Islands Government expedition in August 1975, two or three specimens were found on the south coast at Hawkesbill Bay, and in January 2008 a few were noted at sites on both north and south coasts of the Brac. *A. jatrophae* was not seen on Little Cayman either during six weeks spent on the island in 1975 or on subsequent visits.

Biology
The larva of *A. jatrophae* is known to feed especially on *Bacopa monnieri* (Water Hyssop), *Blechum pyramidatum* (Blechum) and *Lippia* (= *Phyla*) *nodiflora*

White Peacock perching, GC (14.x.2007), KDG

(Match Head), but its food-plant in the Cayman Islands is unknown. When fully grown, the caterpillar is black with bands of white spots and black, branched spines. There are two clubbed horns on the head.

Anartia jatrophae is a low-flying butterfly that can be seen throughout the year. It was the first butterfly to reappear just a few days after Hurricane Ivan devastated Grand Cayman in September 2004. A broad range of plants is visited for nectar on Grand Cayman, both herbaceous plants such as *Bidens alba*, *Lippia nodiflora* and *Stachytarpheta jamaicensis*, and shrubs including *Caesalpinia pulcherrima*, *Colubrina cubensis*, *Duranta erecta*, *Jatropha multifida*, *Lantana camara*, *Lippia alba* and *Vitex agnus-cati*. Males are territorial. The White Peacock occasionally flies to light at night.

Malachite

Siproeta stelenes (Linnaeus, 1758)
Plate II (15)

Recognition

FWL 45-49 mm. This beautiful insect is quite unmistakable. It is the only Caymanian butterfly extensively marked with green. The upper surface has two series of pale green spots on a black ground colour, a submarginal series of rounded spots which is complete on the hindwing but is represented by only two spots on the forewing, and a median band which on the forewing is broken into large blotches. On the underside, the green spots are much paler, silvery green, and the ground colour is pale silvery brown with a few white and red-brown markings. A white streak on the side of the thorax continues into the cell of the forewing. The outer wing margins are scalloped and there is a short hindwing tail.

Malachite, GC (8.ii.2006), RRA

Subspecies

Siproeta stelenes biplagiata Fruhstorfer, 1907 occurs in south Florida (a relatively recent colonist), Mexico, Central America, Cuba, Isle of Pines and Grand Cayman. It is distinguished from the typical subspecies by the presence of two, not one, pale green spots in the forewing cell. The nominate subspecies is found in Jamaica,

Hispaniola, Puerto Rico and through to some of the Lesser Antilles, and in South America. The distinction between the two subspecies, however, is not always clear (Smith *et al.* 1994).

Species' range
Southern Texas and Florida, Central America, West Indies and South America to Brazil.

Cayman Islands distribution
Grand Cayman, Little Cayman, Cayman Brac

Habitat
Butterflies may be found at woodland edges and overgrown hedges, and in parks and large gardens, always in the vicinity of relatively tall trees, although the larval food plants are herbaceous.

Malachite at rest with wings closed, GC (25. viii.2002), AS

History
The Malachite has presumably been continually present on Grand Cayman since its initial discovery in 1938 when it was 'always scarce ... our four specimens ... are the fruits of four and a half months of trying to obtain a series ...' (Carpenter & Lewis 1943). It remained a very rare insect, seemingly confined to the George Town and West Bay areas of Grand Cayman, until about 2000 when sightings began to increase. Its range on Grand Cayman also expanded, and in 2006 eight individuals were seen nectaring together in the Queen Elizabeth II Botanic Park.

S. *stelenes* was first observed in the Sister Islands in 2005 when Frank Roulstone found it on Little Cayman. In January 2008 solitary individuals were seen at the eastern coast of South Sound Hole, Little Cayman and at Spot Bay, Cayman Brac, the latter a new island record.

Biology
The eggs of *Siproeta stelenes* are dark green and laid singly on leaf-tips of seedlings of species of *Blechum*, and less commonly *Ruellia*, both Acanthaceae (Smith *et al.* 1994). A butterfly was observed in a George Town garden on *Blechum pyramidatum*, but the immature stages of S. *stelenes* have not been recognized in Grand Cayman. The dark chocolate to black caterpillar is ringed intersegmentally with red to purple, the prolegs are pink, the black head carries a pair of branched horns, and the thoracic and posterior abdominal segments are adorned with warty, orange protuberances from which branched spines arise. The light green pupa has a pair of head appendages, a spine on the thorax and several spines on the abdomen (Dethier 1940, Scott 1986b).

The flight period of *S. stelenes* extends all year, but most of our sightings have been between December and April. It flies at moderate height and perches most often on tree foliage, usually with wings closed. Butterflies have been observed nectaring at *Coccoloba uvifera*, *Colubrina cubensis*, *Haematoxylum campechianum*, *Petitia domingensis* and other flowering shrubs and trees. They are also strongly attracted to fallen and fermenting fruits such as those of *Artocarpus altilis* (Breadfruit), *Coccoloba uvifera* (Sea Grape) and *Pouteria campechiana* (Egg Fruit).

The Malachite is a butterfly that birds find very palatable. In an experiment in which Malachites, *Anartia jatrophae*, *Agraulis vanillae*, *Danaus plexippus* and five other species were presented to caged Blue Jays and two species of Tanagers, Brower (1984) found that the Malachite was the species most preferred by Blue Jays, followed by *Anartia jatrophae* and *Agraulis vanillae*, with *Danaus plexippus* the most rejected. Results with the two species of Tanager were similar, except that one preferred *Anartia* to *Siproeta* and for both *Agraulis* was lower in the preference order. It may be mentioned that the Malachite became distinctly more numerous after Hurricane Ivan decimated the small bird populations of Grand Cayman in September 2004.

Siproeta stelenes strongly resembles the heliconiid *Philaethria dido* (Linnaeus), and has been suspected of being its Batesian mimic, but the two species occupy different habitats and only partly overlapping ranges. *Heliconius charithonia* (page 65) has been suggested as a more likely but imperfect model (Brown & Heineman 1972, DeVries 1987).

Crescent Spot

Phyciodes phaon (Edwards, 1864)
Plate II (12,13)

Recognition
FWL 13-16 mm. This is a small butterfly with rounded wings and a complicated upper surface pattern of dark brown and cream on a red-brown ground colour. The underside of the forewing is marked much as the upperside, but the hindwing underside is oatmeal colour with fine, dark lines and spots.

Subspecies
None has been described.

Species' range
The range of the Crescent Spot is from the southern United States to Central America south to Guatemala, and in Cuba (since about 1930) and Grand Cayman. It is thought that *Phyciodes phaon* reached the Cayman Islands and Cuba from Florida, the reverse of the usual direction of colonization. The allied North American species *P. tharos* (Drury, 1773) flies with *P. phaon* in Florida. The discal band on the forewing which is cream-coloured in *P. phaon* is red-brown in *P. tharos*, similar to the ground colour of the rest of the wing.

Cayman Islands distribution
Grand Cayman, Little Cayman, Cayman Brac

Habitat
This is a species of open grassy and weedy places, especially where the water-table is high.

History
The Oxford University Biological Expedition

Crescent Spot, GC (17.i.2008), RRA

it rather inconspicuous. Butterflies on Grand Cayman have been observed taking nectar from flowers of *Lippia nodiflora*, a larval food-plant, and *Spilanthes urens*.

Painted Lady

Vanessa cardui (Linnaeus, 1758)
Plate II (14)

Recognition
FWL 28-33 mm. The Painted Lady appears mainly sand-coloured in flight. The upper surface is tinged with pink, the apex of the forewing is broadly black with white spots and the hindwing has a submarginal series of round, black spots as in *Euptoieta hegesia* (below). The under surface of the hindwing is patterned in different shades of brown to cream, with a submarginal series of usually four eye-spots with black and blue pupils circled by yellow and black rings; the first and fourth of these eye-spots are larger than the others. The similar but smaller American Painted Lady (*Vanessa virginiensis* (Drury)) has only two hindwing underside eye-spots, but these are large.

Subspecies
None

Species' range
This species has a claim to be the world's most widely distributed butterfly, 'liable to occur almost anywhere except in arctic conditions' (Riley 1975). In the New World, however, it is known as a breeding resident only in the south-western United States, and a small population is thought to exist in the high mountains of Cuba (Hernández 2004). It is a strongly

to the Cayman Islands in 1938 observed swarms of this species in Grand Cayman, but it was not found on Little Cayman or Cayman Brac (Carpenter & Lewis 1943). It was not until January 2008 that the Crescent Spot was recorded in the Sister Islands (Jackson's Bay and Mary's Bay, Little Cayman, and West End, Cayman Brac).

Biology
In 1938 *Phyciodes phaon* appeared to be associated with the creeping herbaceous plant *Sphagneticola* (= *Wedelia*) *trilobata* (Marigold or Ox-eye), but no juvenile stages were found and the plant was presumably visited for nectar. Scott (1986b) gives an account of eggs, larvae and pupae on *Lippia* (= *Phyla*) *nodiflora* (Match Head or Mat Grass) in the United States. The green eggs are laid in clusters on the under surfaces of *Lippia* leaves. The fully grown larva is dark brown with white speckling, many cream-coloured, branched spines, a black mid-dorsal line and a pair of cream lateral lines. The pupa is red-brown with cream and black mottling.

The Crescent Spot flies close to the ground and its complex pattern renders migratory insect and occurs as a scarce vagrant in Florida, the West Indies and South America.

The primarily North American *Vanessa virginiensis* is also migratory and, although not recorded from the Cayman Islands, has a resident population on Cuba (Hernández 2004) and has been found in all of the Greater Antilles.

Cayman Islands distribution
Grand Cayman (vagrant)

Habitat
Within its breeding range, *Vanessa cardui* is a butterfly of open areas generally, occurring along roadsides, hedgerows and field edges, and on uncultivated land, especially where there are thistles. It is a frequent visitor to suburban gardens in Europe.

History
A worn Painted Lady captured at Barkers on Grand Cayman in November 1987 by Simon Conyers (Askew 1988) is the first reported Caymanian specimen. Another was seen by Ann Stafford in George Town on 8 November 1997.

Biology
The larvae of *V. cardui* feed mainly on species of *Cirsium* (Thistles) in the Old World, but other food-plants such as *Urtica* (Nettles) are sometimes eaten. It is not known whether or not there is any

Painted Lady nectaring at Lavender in France (20.vi.2008), RRA

acceptable food-plant growing in the Cayman Islands.

The butterfly's flight is fast and powerful, as befits a migratory species, but males often occupy territories on small elevations which they return to repeatedly after being alarmed. The Painted Lady seen in 1997 was nectaring at *Lantana camara*.

Mexican Fritillary

Euptoieta hegesia (Cramer, 1779)
Plate II (16-18)

Recognition
FWL 25-34 mm. The upper surface has a brownish orange ground colour with black lines and spots on the forewing and a series of black submarginal spots and marginal lines on the hindwing. The under surface of the hindwing is mottled purple-brown. There are no tails. In its general colouration *Euptoieta hegesia* resembles *Agraulis vanillae* (below), but the black pattern differs in the two species, and the Mexican Fritillary has shorter, broader wings and the underside of the hindwing is without silver marks.

Subspecies
The nominate subspecies, *E. hegesia hegesia,* is found in the Cayman Islands.

Species' range
The Mexican Fritillary is found from Texas through Central America to Argentina, and in the Bahamas, Greater Antilles and associated islands.

Cayman Islands distribution
Grand Cayman, Little Cayman, Cayman Brac

*Female Mexican Fritillary on Dashalong (*Turnera ulmifolia*), GC (18.iii.2004), RRA*

Mexican Fritillary GC (11.ii.2006), RRA

Larva of Mexican Fritillary on Corky Stem Vine (Passiflora suberosa), GC (10.v.2007), AS

Pupa of Mexican Fritillary, CB (30.i.2008), RRA

Habitat

This is a species of both scrubby, uncultivated ground and of parks and gardens, found wherever its favoured food-plant, *Turnera ulmifolia,* grows. *T. ulmifolia* is a showy, bushy herb with yellow flowers. It grows quickly and is often cultivated.

History

The Mexican Fritillary was found abundantly on all three Cayman Islands in 1938 (Carpenter & Lewis 1943), but in 1975 it was rare in Grand Cayman and Cayman Brac, and numerous only on the coastal strip of Little Cayman (Askew 1994). It remained either scarce or overlooked on Grand Cayman until about 1999 when the population seemed to increase rapidly to the level of abundance it now enjoys.

Biology

The most frequently eaten larval food-plant in the Cayman Islands is *Turnera ulmifolia* (Dashalong), and this is often seriously defoliated by the caterpillars. Flower petals as well as leaves are eaten. On Grand Cayman, the Passion Flower vines *Passiflora caerulea, P. cupraea* and *P. suberosa* are also food-plants. The female butterfly lays one egg on a leaf. A fully grown larva is purple-red with three longitudinal creamy-white lines edged with black. There are branched spines on the body, and two forwardly-directed, long, black spines on the front of the thorax. The pupa is dark brown with two silver markings on each side, a c-shaped mark on the front of the wing-case and an oblique bar crossing the middle of the wing-case and continuing down the side of the abdomen. Thorax and abdomen of the pupa bear ten pairs of black-tipped, conical, dorsal tubercles with an additional pair of small tubercles on the sides of some of the abdominal segments. From one pupa collected in 1975 on Grand Cayman there emerged about twenty chalcid wasps which are indistinguishable from European *Pteromalus puparum* (Linnaeus).

The adult butterfly is low-flying and usually feeds at flowers of herbaceous plants, in particular *Turnera ulmifolia* and *Stachytarpheta jamaicensis.* It has also been observed nectaring in Grand Cayman on the shrubs *Caesalpinia pulcherrima, Duranta erecta* and *Jatropha multifida.*

Long-wing Butterflies

Heliconiidae

This is a neotropical family, sometimes treated as a subfamily of Nymphalidae. Heliconiids are brightly coloured butterflies with a distinctive wing-shape, the forewing being long and narrow with its apex extended much further beyond the outer margin of the hindwing than in other families of butterflies.

The larvae of Heliconiidae feed on Passion Flower vines (Passifloraceae). Alkaloids from these plants are sequestered by the larvae so that they, and ensuing stages, are unpalatable to predators. Butterflies which adopt distastefulness as a protection, rather than concealment (crypsis), must advertise that they are to be avoided. They must evolve an appearance, a warning (aposematic) pattern, which predators will remember. An inexperienced or naïve predator will then rapidly learn, after just a few encounters, to associate an unpleasant experience with this warning pattern. The memorability of a warning pattern is enhanced if it is simple, and in contrasting bold colours which are repeated on upper and under wing surfaces. Aposematic butterflies often fly slowly or glide, and they may aggregate, so as to enhance the visual impact of their warning colouration. Heliconiids are resilient butterflies with a capacity to tolerate some physical damage incurred when 'sampled' by a would-be predator. They are also exceptionally long-lived butterflies. The three Caymanian representatives of Heliconiidae together display all of these characteristics.

Gulf Fritillary

Agraulis vanillae (Linnaeus, 1758)
Plate III (1,2)

Male Gulf Fritillary, GC (18.i.2008), RRA

Recognition
FWL 30-38 mm. (specimens from Grand Cayman have FWL usually in excess of 34 mm., those from the Sister Islands are on average smaller). The upper surface of the Gulf Fritillary is orange, brighter in the male, with black markings. At a distance it could be mistaken for *Euptoieta hegesia* (above) or *Dryas iulia* (below), but the silver markings on the under surface of the hindwings readily distinguish *Agraulis vanillae* (which is often placed in the genus *Dione* Hübner, 1819) from both of these species. The underside of the forewing also has some apical silver spots, and there is a pink basal flush in the freshly emerged insect.

Gulf Fritillary GC (9.ii.2006), MLA

*Female Gulf Fritillary captured by a white crab spider on Rabbit Thistle (*Tridax procumbens*), CB (26.i.2008), RRA*

Subspecies

Agraulis vanillae insularis Maynard, 1889 flies in the Bahamas and more northerly West Indies, from Cuba to Dominica and including the Cayman Islands. It is characterised by a well-developed black spot mid-way between base and outer margin in the space between the two hindmost veins of the forewing. This spot is reduced or absent in most individuals belonging to the other subspecies, but butterflies at the indefinite southern extent of the range of *A. v. insularis*, in the Lesser Antilles, intergrade with the nominate subspecies.

Species' range

Agraulis vanillae occurs in the Gulf of Mexico area and Florida (as *A. v. nigrior* Michener), migrating north in the United States but not breeding beyond the northern limit of *Passiflora* survival. *A. v.*
insularis ranges from the Bahamas south to about Dominica (above), and the nominate subspecies (*A. v. vanillae*) extends the range of the species from St Lucia, and perhaps Martinique, into South America as far as Argentina.

Cayman Islands distribution

Grand Cayman, Little Cayman, Cayman Brac

Habitat

The Gulf Fritillary is a common Caymanian butterfly which may be seen almost anywhere in the islands except in mangroves and thick woodland. It is plentiful on beach ridges and in pastures, but perhaps has a preference for scrub, wood edges, hedgerows and uncultivated ground verging paths and roads. It is a familiar garden insect.

History

Agraulis vanillae was recorded as abundant on all three Cayman Islands in both 1938 (Carpenter & Lewis 1943) and 1975 (Askew 1980). This continues to be its status.

Biology

The larval food-plants of the Gulf Fritillary are species of *Passiflora* (Passion Flower vines), over forty species of which are listed as hosts by Beccaloni *et al.* (2008). In the Cayman Islands *P. caerulea*, *P. cupraea* and *P. suberosa* are known to be eaten. Eggs are laid on leaf-tips and tendrils. The fully grown caterpillar is blackish with rust-coloured longitudinal stripes and six longitudinal rows of branched spines arising from black spots on the abdominal segments, and there are two backwardly-curved head spines. Although the larva is supposed to be distasteful, we have seen caterpillars being predated by Assassin Bugs (Reduviidae) and Crab Spiders (Thomisidae).

Adult butterflies can be seen in every month. They fly from early morning to sunset, with greatest activity around midday and no apparent retirement to shade when the sun is at its strongest (Askew 1994). Many plants are used as sources of nectar; we have seen *A. vanillae* at *Bidens alba*, *Bougainvillea*, *Caesalpinia pulcherrima*, *Cissus trifoliata*, *Colubrina cubensis*, *Cordia* species, *Duranta erecta*, *Jatropha multifida*, *Stachytarpheta jamaicensis* and *Waltheria indica*. Unlike *Heliconius charithonia* (below), pollen is not collected by *A. vanillae*, and the female butterfly's rate of egg laying rises to a peak after a few days of adult life, thereafter quickly declining (Dunlap-Pianka *et al.* 1977). The Gulf Fritillary is a somewhat distasteful butterfly to some predators, but a male was observed on Little Cayman attempting to copulate with a female held by a Crab Spider.

Julia, Flambeau

Dryas iulia (Fabricius, 1775)
Plate III (4-6)

Recognition

FWL 37-44 mm. This is a mainly golden orange butterfly, brighter in the male than in the female. The male has a small triangular black mark at the end of the forewing cell, with sometimes a second black spot just distal to it. Occasionally, these two black spots are connected. The wing margins are very narrowly black. The upper surface of the female is more brownish and the black markings are more extensive. A complete black bar runs from the middle of the costal margin of the forewing, where it is broadest, across the end of the cell, and tapers to the outer wing margin. The margin of the hindwing of the female is more broadly black than in the male and it encloses a series of pale spots. The underside is reddish brown in the male, browner in the female, in both sexes with darker marbling and a transverse white streak near the anterior margin of the hindwing.

Seasonal variation in north Caribbean *Dryas iulia* is described by Miller & Steinhauser (1992), the under surface of the wet season (June to November) form being more strongly patterned.

Subspecies

A non-migratory butterfly, no fewer than ten subspecies of *D. iulia* have been described in the Caribbean, each endemic to a different island or island group. The

*Male Julia on Spanish Needle (*Bidens alba*), GC (26.i.2008), TP*

*Julia larva on Wild Red Passionflower (*Passiflora cupraea*), GC (8.iv.2003), AS*

Female Julia, GC (18.i.2008), RRA

Pupa of Julia on Passiflora cupraea, *GC (9.iv.2003), AS*

most recently described is *D. i. zoe* Miller & Steinhauser, 1992 from the Cayman Islands, the type locality being Great Beach on Grand Cayman. Previously, Caymanian butterflies were referred to *D. i. cillene* (Cramer, 1779), a name erroneously bestowed on the Cuban form (Carpenter & Lewis 1943), or to *D. i. carteri* (Riley, 1926), the Bahamian subspecies (Riley 1975). *D. i. zoe* is perhaps most closely allied to *D. i. nudeola* (Bates, 1935) from Cuba.

Species' range
A species found from the southern United States (Texas, Florida) through Central America, the Bahamas, Greater and Lesser Antilles and in South America east of the Andes as far south as Bolivia, Brazil and Uruguay.

Cayman Islands distribution
Grand Cayman, Cayman Brac

Habitat
Dryas iulia is mostly seen in light woodland, bushy places or along hedgerows. Quite large aggregations are sometimes found on flowering shrubs.

History
In 1938 *D. iulia* was found only on Grand Cayman where it was 'quite local in its distribution and nowhere very common' (Carpenter & Lewis 1943). In 1975 it was again found only on Grand Cayman

Group of male Julias attracted to a female, GC (9.iii.2004), RRA

Pair of Julias in copulation, male to the left,
GC (14.i.2004), LB

and there rarely. In 1985 and 1995 it appeared to be more numerous, and since 1997 it has been recorded as a common butterfly on Grand Cayman. It was added to the Cayman Brac species list by Miller & Steinhauser (1992) who described the endemic subspecies *D. i. zoe* from specimens collected in 1990 on Grand Cayman (the holotype) and Cayman Brac.

Biology

The yellow, ribbed eggs are laid on the tips of leaves or tendrils of *Passiflora* species (Passion Flower vines). Only *P. cupraea* is confirmed as a food-plant in the Cayman Islands. The immature stages of *D. iulia* are described by Smith *et al.* (1994). The fully grown caterpillar, which feeds at night, has a reddish head with black patches and two long, curved spines, and a blackish body with red-brown legs, ventral surface and last two segments. There are six longitudinal rows of black, branched spines extending over most of the body. The brownish pupa has golden spots on the thorax, and white spots ventrally behind the wing-cases, and there are segmentally arranged pairs of projections, those on the thorax the longest.

Female *D. iulia*, although relatively large butterflies, are rather inconspicuous as they fly low and slowly with shallow but rapid wing-beats, threading their way through dense tangles of vegetation at the bases of shrubs and trees. The more brightly coloured males are more frequently seen. They have been observed nectaring at flowers of *Asclepias curassavica*, *Caesalpinia pulcherrima*, *Chromolaena odorata* and *Jatropha multifida* on Grand Cayman. Elsewhere, *D. iulia* is reported to feed at cow pats (Smith *et al.* 1994).

Zebra

Heliconius charithonia (Linnaeus, 1767)
Plate III (3)

Recognition
FWL 38-46 mm. The elongated black wings with pale yellow bands, three on the forewing and one on the hindwing, plus two series of yellow spots on the hindwing, make the Zebra an unmistakable butterfly.

Subspecies
Caymanian insects are usually attributed to the Cuban subspecies *Heliconius charithonia ramsdeni* Comstock & Brown,

1950 (Miller & Steinhauser 1992), but Miller *et al.* (1994) remark that it is uncertain whether Cayman Islands insects are more like *H. c. ramsdeni* or the Jamaican *H. c. simulator* Röber, 1921. Subspecific differentiation of *H. charithonia* is not clear-cut.

Species' range
Heliconius charithonia is found from the southern United States, through Central America, the Bahamas, Greater and Lesser Antilles, and into South America as far south as Ecuador and Bolivia. It breeds further north than any other species of *Heliconius*.

Cayman Islands distribution
Grand Cayman, Little Cayman, Cayman Brac

Habitat
The Zebra is essentially a woodland butterfly in the Cayman Islands, but it may also be seen in parks, large gardens and along roads and tracks edged by tall trees. It seems to require some degree of shade.

History
In 1938 *H. charithonia* was found only

Zebra, GC (21.i.2008), RRA

*Zebra nectaring on Bull Hoof (*Bauhinia divaricata*), GC (21.i.2008), MLA*

on Grand Cayman but 'abundantly, in south-eastern portions ... especially in the vicinity of Georgetown' (Carpenter & Lewis 1943). In 1975 it was seen on all three Cayman Islands but its status was either rare or local. On Grand Cayman since 1985 the Zebra has always been of local distribution, and it has varied in abundance from being described as scarce to quite common. On Little Cayman in 2007 and 2008 it was abundant.

Biology

The larval food-plants of the Zebra are species of *Passiflora* (Passion Flower vines), but the immature stages have not yet been found in the Cayman Islands. The fully grown larva is white with black spots and longitudinal rows of long, black, branched spines. The head bears a pair of 'long, knobbed coronal setae' (Smith *et al.* 1994). The pupa is yellow-brown

with silver spots and black marks, and it too has a pair of long, curved serrate processes on the head.

Male butterflies are attracted to female pupae, presumably by pheromones, and mating can take place by the male actually penetrating the pupa before the female has completely emerged (Gilbert 1984). Female Zebras produce large eggs at a relatively low but steady rate throughout their unusually long lives which can last several months. They mate repeatedly.

H. charithonia is usually seen on the wing, flying quite slowly but buoyantly with shallow wing-beats at a height of one to three metres. The butterflies take both nectar and pollen from flowers. More pollen is eaten by females than males, the protein obtained making a substantial contribution to their egg production and longevity. On Grand Cayman Zebras have been noted visiting *Bidens alba*, *Caesalpinia pulcherrima* and *Jatropha multifida*. An unusual habit of the species is its gathering before sunset into roosting aggregations (Waller & Gilbert 1982). These communal roosts are on selected branches, usually dead ones devoid of foliage, and the same roosting place may be used by successive generations over many months. The butterflies hang from twigs with wings tightly closed.

Hairstreaks and Blue Butterflies

Lycaenidae

Lycaenidae is a large, worldwide family of small butterflies with over six thousand described species. This is about one-third of the total of all known butterflies. Ten species have been recorded from the Cayman Islands, and these represent the two major subfamilies, Theclinae (Hairstreaks) and Polyommatinae (Blues). In Blues there is generally considerable sexual dimorphism in wing colour, but the sexes of Hairstreaks are usually rather similar. Caymanian Theclinae, with the exception of *Eumaeus atala,* have one or two tails on each hindwing, but the four Caymanian species of Polyommatinae have at most only rudiments of tails.

Polyommatinae are low-flying butterflies found mostly in open, grassy and flowery situations. Very few show any tendency to migrate, so that island populations can be long isolated and hence are often distinctive (Nabokov 1945). The great majority use herbaceous Fabaceae (Leguminosae) as larval food-plants, and their slow-moving, woodlouse-shaped caterpillars often feed in the flower heads. Blue Butterfly larvae are frequently attended by ants in a symbiotic relationship in which the caterpillar receives some protection from arthropod predators and probably parasitic wasps, and the ant in return receives a sweet, sugary secretion from glands on the caterpillar (Pierce 1984).

Theclinae tend to be more associated with wooded areas and bushy places, often flying about the tops of trees, but this is not true of *Strymon istapa*, the commonest of the Cayman Hairstreaks.

Pupae in both subfamilies of Lycaenidae are dumpy and have a silken girdle which usually holds the ventral surface close to the substrate, unlike pupae in the preceding families which are suspended head downwards from a silk pad by a hooked organ (cremaster) at their posterior ends.

Adult Lycaenidae have six functional legs in females, but in males the front pair are reduced and not used for walking. They usually perch with wings closed, and the settled butterfly will slowly raise and lower its hindwings against the forewings. Such movement may enhance the head-like appearance of the tornal area on the under side of the hindwing, with its eye-spot markings and antenna-like tails. This illusory head is thought to attract a hunting bird or lizard, so that its attack is drawn to the relatively expendable hindwing tornal area and deflected from more vital body parts.

Atala Hairstreak

Eumaeus atala (Poey, 1832)
Plate III (7-9)

Recognition

FWL 17-22 mm. (males smaller than females). Unlike other Hairsteaks, the wings of *Eumaeus atala* are rounded, and there are no hindwing tails. The upper surface has a black ground colour with blue (male) to green-blue (female) iridescent scaling on the forewing except on the veins and broad margins; there are basal patches of the same colours on the hindwings which also bear a marginal series of green-blue spots. The under surface is similar in both sexes, the forewing entirely black, the hindwing with three arcs of iridescent bluish spots in the distal half and a bright red spot at the centre of the inner margin. The abdomen is red. Altogether the Atala is a striking butterfly which is unlikely to be mistaken for any other Caymanian insect except perhaps, at a distance, the Faithful Beauty moth (page 159).

Subspecies
Only the nominate subspecies *Eumaeus atala atala* occurs in the Caribbean. Populations in Florida were named as *E. a. florida* Röber, 1926, but this is poorly differentiated from the typical form.

Species' range
The species is restricted to Florida, the Bahamas, Cuba and the Isle of Pines, and Cayman Brac. In Florida it is a very local butterfly and appears to exist in rather temporary colonies in which the butterflies may be plentiful for a time and then die out.

Reared female Atala, CB (30.i.2008), RRA

Cayman Islands distribution
Cayman Brac

Habitat
Eumaeus atala is nearly always found close to cycads (*Zamia*) on which the larvae feed. On Cayman Brac it is most numerous in dry woodland on the bluff. *Zamia* grows on all three Cayman Islands, but to date the Atala has been seen only on Cayman Brac.

History
The Atala Hairstreak was discovered in large numbers on Cayman Brac in November, 1990 (Miller & Steinhauser 1992). It was found at Stake Bay on the north coast, Hawkesbill Bay on the south coast, and was also seen at Brac Reef Resort. It now seems to occur wherever its food-plant grows, but the main population appears to be on the central bluff.

Male Atala, CB (28.i.2008), RRA

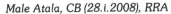

Pair of Atalas in copulation, female partly hidden by leaf, CB (28.xii.2007), WP

Biology

The predominant red on black colouration of the adult butterfly, together with its slow flight, signifies that it is distasteful. Similarly the gregarious larva, brownish red with 7 pairs of slightly-raised, dorsal, transverse, pale yellow tubercles surmounted by short, black bristles, seems to be warningly coloured. They are not attended by ants. The pupa also is conspicuous, red-brown with faint yellow spots, usually found in groups on vegetation close to the larval food-plant. The gregarious nature of all life-history stages suggests that they are distasteful; other pre-adult Lycaenidae are usually solitary. Toxic alkaloids are probably stored by the caterpillars as they feed on cycads. *Zamia integrifolia* (= *pumila*) (Bull Rush or Coontie) is the larval food-plant on Cayman Brac (and Cuba). Eggs are laid in batches of usually 10-20 on the under surfaces of young leaves of *Zamia,* and the larvae may be sufficiently numerous to defoliate the food-plant. *Zamia* growing

*Atala larvae on Bull Rush (*Zamia integrifolia*), CB (28.i.2008), MLA*

Atala pupae, CB (29.i.2008), RRA

in open, unshaded situations appears to be avoided.

In the middle of the day adult butterflies are usually encountered singly, flying quite low in the woodland scrub, but two hours or so before sunset they begin to congregate about favoured trees, chasing one another and occasionally resting on sunlit leaves. Whether these congregations are composed mostly of males or of both sexes is not known.

Antillean Hairstreak

Chlorostrymon maesites (Herrich-Schäffer, 1864)
Plate III (10)

Recognition
FWL 10 mm. The upper surface of the male is brilliant purple-blue, that of the female mainly blue, but the underside in both sexes is predominantly bright green with a broken postdiscal line on both wings. The hindwing, when intact, bears two tails.

Subspecies
The single known Caymanian specimen belongs to the nominate subspecies.

This differs from *Chlorostrymon clenchi* (Comstock & Huntington, 1943), at one time considered a subspecies of *C. maesites*, in having two, not one, hindwing tails.

Species' range
Chlorostrymon maesites occurs in southern Florida, the Bahamas, Greater Antilles and some of the Virgin Islands and Lesser Antilles. *C. clenchi* flies only in Dominica, and *C. telea* Hewitson has a continental distribution in South and Central America, occasionally reaching as far north as southern Texas.

Cayman Islands distribution
Grand Cayman (vagrant)

Male Antillean Hairstreak from mosquito trap, the only known Cayman specimen, GC (10.vii.2003), JR

Habitat
In Jamaica, the Antillean Hairstreak is described by Brown & Heineman (1972) as 'a butterfly of open ground, where it is found in fields that have begun to be overgrown. Its flashing colors make it easy to follow as it flits from bush to bush.'

History
As far as is known, only a single specimen of this distinctive little Hairstreak has

Under surface of male Antillean Hairstreak from mosquito trap, GC (10.vii.2003), JR

been found in the Cayman Islands. On 10 July, 2003 a worn but unmistakable male specimen was found by David Malone and Joanne Ross in the catch of a light trap operated by the Mosquito Research and Control Unit on the Esterly Tibbets Highway (Harquail Bypass) north of George Town, Grand Cayman.

Biology
The larval food-plant is *Cardiospermum* (Sapindaceae), two species of which are native to the Cayman Islands.

Cuban Grey Hairstreak

Strymon martialis (Herrich-Schäffer, 1865)
Plate III (11,12)

Recognition
FWL 12-14 mm. The upper surface is dark brown with a broad, blue band on the inner margin of the forewing and blue scaling over the posterior three-quarters of the hindwing. The male has a round, dark patch of androconial scales at the apex of the forewing cell. There are two thin tails on the hindwing, the posterior one the longer, a small orange spot at the tornus, and a small black marginal spot between the bases of the tails. The underside is grey with a white line, edged internally with black, running across both fore- and hindwings, and a yellow-orange mark, enclosing a black spot, near the bases of the tails. There is also a dark, red-brown spot at the hindwing tornus, and a small, blue spot between this and the yellow-orange mark. Under surface markings are similar to those of *Strymon acis* (below) but there is just a single white line running to the yellow-orange mark at the base of the tails, and there are no small white spots on the hindwing between the white line and wing-base. In *S. acis*, an additional submarginal white line joins the orange mark at the bases of the tails, and the orange mark encloses only a vestigial black spot.

Subspecies
None

Species' range
Strymon martialis is found in Florida, the Bahamas, Cuba and the Isle of Pines, the Cayman Islands and Jamaica.

Cayman Islands distribution
Grand Cayman, Little Cayman

Habitat
This Hairstreak is not uncommon in Little Cayman where it frequents the beach ridges, males perching on prominent twigs of *Conocarpus erectus* or of the larval food-plant *Suriana maritima,* apparently holding territories.

History
Carpenter & Lewis (1943) and Askew (1980) record *S. martialis* only from Little

Cayman in 1938 and 1975 respectively. The first Grand Cayman specimens (two) were seen in November 1985 at Boatswain Bay (Schwartz *et al.* 1987), and a further two were found at Seven Mile Beach in November 1990 (Miller & Steinhauser 1992). More recently on Grand Cayman, single butterflies have been observed in October 1995 (Half Way Pond), August 1997 (George Town), March 2004 (Cottage), February 2006 (Old Man Bay), December 2006 (South Sound) and February 2008 (Cottage). This history of *S. martialis* indicates that it is a rare butterfly on Grand Cayman, in contrast to its status on Little Cayman where, in 1975, it was 'Frequently seen about *Conocarpus* at Pirates Point (Preston Bay), and also recorded at Blossom Village on the storm beach, north shore path, Owen Island, and in the interior south of Crawl Bay' (Askew 1980). In 2008 a number of specimens were seen on Little Cayman, especially on the beach ridges of the south coast.

Biology

A larval food-plant of *S. martialis* is *Suriana maritima* (Jennifer or Juniper) (Smith *et al.* 1994), which is a coastal

Male Cuban Grey Hairstreak on Sea Lavender (Argusia gnaphalodes), LC (23.i.2008), RRA

Cuban Grey Hairstreak, GC (12.xii.2006), JG

shrub, and *Trema lamarckianum* (Trema) may also be eaten. Both of these plants grow in the Cayman Islands but immature stages have yet to be found there. The larva is dull green with faint yellow lines and a dense covering of short white hairs.

In common with other Hairstreaks, the butterflies generally perch with wings closed and slowly move their hindwings in time against their forewings creating the illusion of a head with eyes (the complex tornal markings) and antennae (the tails). A predator such as a lizard or bird hunting by sight may be decoyed into attacking this false head, damage to which would not seriously harm the butterfly. The butterfly has been observed feeding at flowers of *Stachytarpheta jamaicensis*.

Drury's Hairstreak

Strymon acis (Drury, 1773)
Plate III (13,14)

Recognition
FWL 12-14 mm. The upper surface is dark grey-brown without blue scaling,

and on the hindwing there is a blackish spot near the base of each tail and a small orange spot and short white line in the tornal area. The male has a dark brown androconial patch at the end of the forewing cell, the only noticeable difference between the sexes. On the under surface, *Strymon acis* is not unlike *S. martialis* (above), having grey-brown ground colour with a strong white line, edged on its inner side with black, across both wings, and on the hindwing a conspicuous orange-red spot at the base of the tails and a black spot at the tornus. *S. acis* may be distinguished from *S. martialis* on hindwing under surface pattern by having a second white line running forwards submarginally from the large orange-red mark to reach the costal wing margin and, most especially, by the presence of two small white spots midway between white line and wing base. In addition, *S. acis* has a larger tuft of reddish hairs on top of its head between the eyes.

Subspecies
Caymanian Drury's Hairstreak is not assigned to a subspecies. A large number

*Drury's Hairstreak on Sea Lavender (*Argusia gnaphalodes*), LC (21.i.2008), RRA*

Drury's Hairstreak nectaring on Rosemary (Croton linearis), LC (23.i.2008), MLA

of more or less distinct island forms have developed in Caribbean populations of *S. acis*, and many of these have been formally described as subspecies. The taxonomy of Drury's Hairstreak in the Cayman Islands is, however, very confused. In 1938 it was noted that Grand Cayman butterflies had two well-developed orange-red spots on the upper surface of the hindwing, one at the tornus, the other separated from the shorter tail by a black spot. In specimens from Little Cayman and Cayman Brac, these orange-red spots were much reduced or even almost absent (Carpenter & Lewis 1943). Riley (1975) refers all Caymanian insects to the Jamaican subspecies *S. acis gossei* (Comstock & Huntington, 1943), and illustrates a male from Cayman Brac. Schwartz *et al.* (1987) attributed only Grand Cayman insects to *S. a. gossei*, and named insects from the Sister Islands as *S. a. casasi* (Comstock & Huntington, 1943), the Cuban subspecies. Miller & Steinhauser (1992), however, took a different view and assigned

a specimen from Cayman Brac and another from Grand Cayman to *S. a. gossei*. Finally, Smith *et al.* (1994) write that *S. a. casasi* 'is apparently the subspecies that flies in the Cayman Islands'. It is improbable that two subspecies occur on the same island, and the difficulty in finding agreement on the naming of Cayman Islands *S. acis* creates doubts about the value of a subspecific epithet. Until the question has been more fully investigated, we adopt the position of Lamas (2004) in not assigning the various island forms of *S. acis* to subspecies.

Species' range

Drury's Hairstreak is more widely distributed than the preceding species, ranging from Florida, the Bahamas and through the Greater and Lesser Antilles as far south as Dominica.

Cayman Islands distribution

Grand Cayman, Little Cayman, Cayman Brac

Habitat

Strymon acis flies with *S. martialis* on the beach ridges of Little Cayman, but it is also found in dry localities inland. In Florida this is a woodland butterfly.

History

Discovered locally on all three Cayman Islands in 1938 (Carpenter & Lewis 1943), Drury's Hairstreak is a rather difficult butterfly to find. It appears to be generally uncommon, although more numerous on Little Cayman, and perhaps also on Cayman Brac, than on Grand Cayman. On Grand Cayman it has been unseen for quite long periods, but it is unclear whether this is because of actual scarcity or the elusive nature of the butterfly.

Biology

A larval food-plant of *S. acis* is *Croton linearis* (Rosemary), but immature stages have not been found in the Cayman Islands. The butterflies have a rapid, darting flight, but they perch frequently and males often return repeatedly to a prominent position on a bush, seemingly holding a territory. Flowering shrubs such as *Croton linearis*, but also herbaceous flowers, provide the insects with nectar.

Dotted Hairstreak

Strymon istapa (Reakirt, 1867)
Plate III (15-17)

Recognition

FWL 11-13 mm. *Strymon istapa* differs from the two preceding species of *Strymon* in having only a single, relatively short, hindwing tail and an under surface without white lines. The upper surface is dark brown in the female with some

Male Dotted Hairstreak on Sea Lavender (Argusia gnaphalodes), LC (23.i.2008), RRA

slightly shining, grey-blue scaling on the basal part of the forewing and posterior part of the hindwing. The male upper surface is dark brown without grey-blue scaling, and has a dark androconial patch in the forewing cell. In both sexes there is a black spot, flanked by two smaller, fainter dark spots, near the base of the hindwing tail. On the grey-brown underside in both sexes, the white lines edged with black that distinguish *S. martialis* and *S. acis* (above) are replaced by series of closely approximated small dark spots and white spots. There is a black spot near the base of the tail, which is capped by a larger orange-red mark, and a small black spot at the tornus.

Subspecies

Strymon istapa from the west Caribbean appears in much of the literature as a subspecies of *S. columella*, *S. c. cybira* (Hewitson, 1874), but *S. columella* is a distinct species found in the Lesser Antilles.

Species' range

S. istapa flies in the southern United States, Mexico, the Bahamas, Cuba and the Isle of Pines, Cayman Islands, Jamaica, Hispaniola and Puerto Rico.

Cayman Islands distribution

Grand Cayman, Little Cayman, Cayman Brac

Habitat

This is a species of road and path edges, uncultivated corners of gardens, waste ground and flowery, disturbed land generally. According to Lewis, *S. istapa* (as *S. columella*) was 'always found [on Grand Cayman in 1938] at the tops of beaches where *Suriana* … is found' (Carpenter &

Dotted Hairstreak on Rabbit Thistle, CB (25.i.2008), RRA

indica (Buff Coat) on Grand Cayman. Several larvae were found in maturing seed-heads of this shrubby herb growing in a derelict garden on the north-east coast in February 1997. The larva is cryptically coloured, yellow-green with a darker, mid-dorsal band and densely covered with short, pale hairs. In captivity, these caterpillars pupated on the upper surfaces of the leaves of the food-plant and butterflies emerged some eight days later. A solitary parasitic wasp (Braconidae) emerged from one pupa.

The butterfly usually flies low and is perhaps more associated with herbaceous plants than are the other Caymanian Hairstreaks. It is often to be found prominently perched on a dead flower-head or flitting about low bushes before suddenly alighting. The Dotted Hairstreak has been noted to take nectar at flowers of *Bidens alba*, *Heliotropium angiospermum*, *Jatropha integerrima*, *Lippia nodiflora* and, in particular, *Spilanthes urens*.

Lewis 1943), and Askew (1980) reported it to be entirely coastal on Little Cayman, flying about *Conocarpus erecta* on the beach ridges.

History
Strymon istapa was discovered in Grand Cayman in 1938 (Carpenter & Lewis 1943), Little Cayman in 1975 (Askew 1980) and Cayman Brac in 1985 (Schwartz *et al.* 1987) (all citations as *S. columella*). In contrast to the two preceding species of *Strymon*, it seems to be more numerous on Grand Cayman than in the Sister Islands, although in 2008 it was abundant on all three Cayman Islands.

Biology
Following Carpenter & Lewis (1943), Smith *et al.* (1994) suggest *Suriana maritima* (Jennifer, Juniper) as a larval foodplant in the Cayman Islands, but we have found caterpillars eating *Waltheria*

Fulvous Hairstreak

Electrostrymon angelia (Hewitson, 1874)
Plate III (18)

Recognition
FWL 10-12 mm. This is a small Hairstreak with two hindwing tails and quite sharply pointed forewings. It is the only Hairstreak recorded from the Cayman Islands to have areas of brownish orange or fulvous colouration on the dark brown upper surface, most extensive on the forewings and in the males. The under surface is brownish, the outer half of the forewing crossed by two fine, dark lines without white edging, the hindwing with an irregular, broken, black line outwardly

The first example of the Fulvous Hairstreak to be found in the Cayman Islands, GC (13.viii.1985)

The first record of this species in the Cayman Islands is of a male collected on 13 August 1985 in George Town, Grand Cayman (Askew 1988). It was flying along a tall hedge bordering the sports ground; the hedge has now been removed. In November 1985 another male was captured on Grand Cayman (locality not stated) (Schwartz *et al.* 1987), and a female was taken at Stake Bay, Cayman Brac on 19 February 1994 by Peter Davey. The female is now in the National Trust reference collection. In addition to the above records, we have received verbal information from Joanne Ross of the occurrence of *E. angelia* in her George Town garden (date unspecified), and from Frank Roulstone who saw 'several' specimens near his home at North Side, Grand Cayman over a period in 2005.

edged with white. The hindwing also has a black spot at the tornus, and another adjacent to a large, triangular orange mark between the bases of the tails. There are small patches of blue and white scales in the tornal area.

Subspecies

Caymanian material is assigned to the nominate subspecies which is found in southern Florida, Cuba and the Isle of Pines. Three other named subspecies occur respectively in the Bahamas, Jamaica and from Hispaniola to the Virgin Islands.

Species' range

The Fulvous Hairstreak is distributed from Florida (colonized in the 1970s), the Bahamas, Cuba, Cayman Islands, Jamaica, Hispaniola and Puerto Rico to the Virgin Islands.

Cayman Islands distribution

Grand Cayman, Cayman Brac

Habitat

Electrostrymon angelia frequently flies about shrubs and trees, but it also nectars at herbaceous plants (Smith *et al.* 1994).

Biology

Reported larval food-plants of the Fulvous Hairstreak are *Schinus terebinthifolius* (Brazilian Pepper) (Smith *et al.* 1994), *Piscidia piscipula* (Dogwood, Fishpoison Tree) and species of *Salvia*. *Schinus* is not native to the Cayman Islands, but young plants were found in 2002 growing in a George Town garden, since built over. The tree is highly invasive, but because it is not salt tolerant, it is unlikely to become widespread in the islands. *Piscidia* is a native but endangered Cayman plant.

The butterfly perches with wings closed on the upper surfaces of leaves, and it roosts in the tree canopy. It takes nectar from ground-layer flowers such as *Bidens* and *Lantana* (Smith *et al.* 1994).

Pygmy Blue

Brephidium exilis (Boisduval, 1852)

Plate III (19,20)

Recognition

FWL 6.5-8.5 mm. This tiny butterfly is the smallest in the Western Hemisphere, possibly in the world. It has a dark brown upper surface with coppery tints and some blue scaling, most evident in the male, at the bases of the wings. The most conspicuous feature of the grey-brown under surface is on the hindwing where there is a submarginal series of four or five black eye-spots with minute silver pupils.

Subspecies

Brephidium exilis thompsoni Carpenter & Lewis, 1943 is known only from Grand Cayman where it was discovered on 23 June, 1938 by the Oxford University Biological Expedition. Carpenter &

Pygmy Blue, GC (6.ii.2008), KDG

Lewis (1943) write 'The tiny butterfly is indeed limited in its distribution for it was not found outside of an area of about fifty square yards, on the edge of a secluded lagoon, known as English Sound, lying to the east of and off of the Great Sound'. The subspecies is named after Gerald H. Thompson, a member of the 1938 expedition, who was responsible for its discovery.

A second subspecies, *B. e. isophthalmia* Herrich-Schäffer, 1862 is found in the Bahamas, Cuba and the Isle of Pines, Jamaica and Hispaniola, whilst the nominate form is confined to the American mainland.

Species' range

The Pygmy Blue occurs from Oregon and Nebraska, through the southern United States and Central America, into northern South America, and in the Bahamas, Cuba, Grand Cayman, Jamaica and Hispaniola.

*Female Pygmy Blue on Glasswort (*Salicornia perennis*), the larval food-plant, GC (27.i.2006), MLA*

Cayman Islands distribution
Grand Cayman

Habitat
The endemic subspecies of the Pygmy Blue is restricted to low-lying, saline situations where the dominant vegetation is *Sesuvium portulacastrum* and *Salicornia perennis*. Such conditions are localized and colonies of *B. e. thompsoni* usually occupy areas of less than one hundred square metres.

History
The type locality, English Sound, is not named as such on recent maps of Grand Cayman, but we believe it to be the small eastern lagoon north of Little Sound, which in turn is an eastern extension of North Sound. It is adjacent to Bowse Land, but we have not been able to reach the lagoon overland. Auburn Myles, cook to the 1938 Oxford University expedition, has told us that Lewis and Thompson were based on a schooner, the Meritwell, in North Sound and they must surely have taken a marine route to English Sound. English Sound was named after T.M. Savage English, a naturalist who resided at North Side from 1912 to 1914 and made observations at Rum Point and neighbouring areas.

After its discovery in 1938, *B. e. thompsoni* eluded several searches before being rediscovered in November 1985 by Simon Conyers (personal communication) who found two widely separated colonies on Grand Cayman, one in the Barkers area and one in the north of the island south of Hutland. In 1995 we were informed of a colony near Bodden Town (Fred Burton, personal communication), and in March 2002 a strong colony was located at Midland Acres. The Pygmy Blue is evidently quite widespread on Grand Cayman, at least in the western half of the island, but its specialized habitat requirements and localized colonies make it an insect very vulnerable to any land disturbance. We know of one colony that has been destroyed recently by housing development. At the time of writing, *B. e. thompsoni* seems to be most strongly established in the dykes area north of George Town (especially beside Uncle Luke's Pond) and in the Midland Acres area. Also, seventy years after first being found on Grand Cayman, Cayman Pygmy Blues were rediscovered at their type locality, near Savage English's old dock, on 18 June 2008 by Tom Watling who reached the shore of English Sound by kayak.

Biology
The genus *Brephidium* includes just three species, one African and two in the New World. This pattern is indicative of a relict

Pygmy Blue resting on Glasswort, GC (6.ii.2008), KDG

distribution of an ancient group of butterflies (Smith *et al.* 1994).

Although immature stages of the Pygmy Blue have not been seen on Grand Cayman, we can confidently say that a larval food-plant is *Salicornia perennis* (Glasswort). This plant is always found growing in the butterfly colonies and we have frequently seen females flying around it, walking over it, or simply resting on it with wings closed. *Batis maritima* (Saltwort) may be an alternative food-plant on Grand Cayman. Male Pygmy Blues flit about *Salicornia* and *Sesuvium portulacastrum*, and both sexes nectar at flowers of these plants and of *Blutaparon vermiculare*. Flight activity does not seem to begin before mid-morning.

The Pygmy Blue was little affected by Hurricane Ivan, fifteen to twenty butterflies being seen in a colony at Uncle Luke's Pond on 29 October 2004, some seven weeks after Hurricane Ivan struck Grand Cayman.

The dragonfly *Erythemis vesiculosa* has been seen chasing Pygmy Blues but the butterflies evaded capture.

Cassius Blue

Leptotes cassius (Cramer, 1775)
Plate III (21-23)

Recognition
Males of this and the two following species are similar and impossible to distinguish with certainty in flight, their upper surfaces being violet-blue with only very small, dark marginal markings. It is the under surface patterns, visible on the resting butterfly and similar in both sexes, that provide the diagnostic characters.

FWL 10-12 mm. The sexes are dissimilar as in other Polyommatinae. The female *Leptotes cassius* is the only Caymanian species of blue butterfly to have a mostly white upper surface, the forewing blue only basally with broad, blackish costal and outer margins and short, transverse, blackish bars. On the hindwing upper surface in both sexes, in the tornal area, there are two small black spots (ocelli), the outermost the larger. These ocelli are better developed on the underside of the hindwing where they include some blue scales, and the larger is narrowly ringed orange. Both wings on the under surface are marked with brown transverse bars, not spots as in the two following species, and the whitish ground colour is extensive.

Subspecies
Butterflies in the Cayman Islands, and also Florida, the Bahamas, Greater Antilles and Mona Island, belong to the northern subspecies *Leptotes cassius theonus* (Lucas, 1857). Two other subspecies are found further south and east in the West Indies, and the nominate form is a South American butterfly. The assertion that specimens from the Cayman Islands are close to *L. c. catilina* (Fabricius, 1793) from the Virgin Islands (Brown & Heineman 1972) is based on a misidentification of *Cyclargus ammon* (see under History).

Species' range
Leptotes cassius occurs from the south of North America, through Central America and the Caribbean, and in South America south to Argentina.

Cayman Islands distribution
Grand Cayman, Little Cayman, Cayman Brac

Habitat

The Cassius Blue usually flies in open, sunny places, but it enters woodland more readily than the other Cayman Islands Blues and here 'can be seen fluttering about in the crowns of the trees, especially of those in bloom' (Brown & Heineman 1972).

History

Carpenter & Lewis (1943) and Askew (1980) reported finding *L. c. theonus* on all three Cayman Islands in 1938 and 1975 respectively. In addition, Carpenter & Lewis record '*Hemiargus catalina* (Fabricius)' from Little Cayman and Cayman Brac, but this refers to *Cyclargus ammon* (Lucas) (below) and not *L. c. catilina* (Fabricius). The Cassius Blue never seems to have been an abundant butterfly in the Cayman Islands and in some years on Grand Cayman it is scarce; in 1985 it was not seen at all and prematurely adjudged to be possibly extinct (Askew 1988). Miller & Steinhauser (1992) encountered *L. cassius* at one site on Grand Cayman and three sites on Cayman Brac in 1990, and records strongly suggest that the species is more numerous in the Sister Islands than on Grand Cayman. This was certainly the case in 2008.

Biology

Leptotes cassius has a broad range of recorded larval food-plants, mainly herbaceous Fabaceae but also Plumbaginaceae. In the Cayman Islands one likely larval food-plant is *Galactia striata* (Milk Pea);

Cassius Blue, GC (18.i.2008), RRA

caterpillars thought to be of the Cassius Blue have been found in its pods.

Like the two following species of Blue, *L. cassius* is a sun-loving insect and may be seen fluttering about and nectaring at flowers of herbaceous plants such as *Bidens alba* and *Stachytarpheta jamaicensis*. It will also visit blossoming shrubs and trees, and is especially attracted to flowers of *Bauhinia divaricata* (Brown & Heineman 1972) and *Haematoxylum campechianum*. On Grand Cayman at the southern end of the Mastic Trail, *L. cassius* has several times been seen flying at a height of two to three metres about bushes over which *Galactia striata* was growing. These butterflies mostly flew rapidly and erratically, but a copulating pair was observed perched high on foliage.

Male Hanno Blue, GC (8.ii.2006), MLA

Hanno Blue

Hemiargus hanno (Stoll, 1790)
Plate III (24-26)

Recognition
FWL 9-11 mm., on average smaller than the Cassius and Lucas's Blues. The male *Hemiargus hanno* is violet-blue above with narrow, black wing borders, and the female is dark brown with blue scaling (purer blue and more shining than in the male) at the base of all wings. The hindwings of both sexes, on upper and under surfaces, have single, small, black marginal spots in the tornal areas. The under surface is grey (not white as in the Cassius Blue) with blackish spots, the largest of which is the tornal spot on the hindwing. *Cyclargus ammon* (below), which might be mistaken for *H. hanno*, can be readily distinguished by having twin hindwing tornal spots.

Subspecies
Hemiargus hanno filenus (Poey, 1832) is found in the Bahamas, Turks and Caicos Islands, Cuba and the Isle of Pines, and the Cayman Islands. It sometimes appears in the literature as *Cyclargus ceraunus filenus*.

Species' range
The range of the Hanno Blue extends from the southern United States, through the Bahamas and West Indies, and through Central and South America south to Argentina.

Cayman Islands distribution
Grand Cayman, Little Cayman, Cayman Brac

Habitat
Hemiargus hanno inhabits beach ridges, gardens, parks, roadside verges, woodland edges and other sunny, open places with secondary vegetation.

*Hanno Blue resting on dead flower-head of Spanish Needle (*Bidens alba*), GC (17.i.2008), RRA*

History

In 1938 *H. hanno* was found only on Grand Cayman (Carpenter & Lewis 1943), in 1975 it was seen on Grand Cayman and Little Cayman (Askew 1980), and in 1985 a single specimen was found at the airport on Cayman Brac (Schwartz *et al.* 1987). It is always a plentiful butterfly on Grand Cayman, but is apparently less common on the Sister Islands, although there are specimens in the National Trust that were collected in 1994 on Cayman Brac by Peter Davey, and it was common both here and on Little Cayman in January 2008.

Biology

Several plants of the family Fabaceae have been recorded as larval food-plants of *H. hanno*. Its diet on the Cayman Islands is not known, but *Crotalaria* species (Sweet Peas), *Abrus precatorius* (Licorice), *Chamaecrista nictitans* (Wild Shame-face), *Mimosa pudica* (Shame-face), *Phaseolus lathyroides* (Wild Dolly) and *Senna alata* (Candle Bush) are likely candidates. Adult butterflies have been noted nectaring at *Ambrosia hispida* on Little Cayman, and on *Bidens alba* and *Spilanthes urens* on Grand Cayman. Their flight is low and weak. At night, like many Blues, they tend to roost in groups, several towards the tops of stems of herbaceous plants and often on dead flower-heads, all with heads directed downwards.

Lucas's Blue

Cyclargus ammon (Lucas, 1857)
Plate III (27-29)

Recognition

FWL 10-12 mm. Lucas's Blue was at one time placed in *Hemiargus* with *H. hanno* (above) which it resembles, but it may be distinguished by the presence, on both the upper and under wing surfaces, of two (not one) blackish spots in the hindwing tornal area. The outermost of these spots is capped with orange on the under surface in both sexes, but on the upper surface orange scaling is conspicuous only in the female and reduced or absent in the male. The grey under surface of *Cyclargus ammon* is more strongly marked than in *H. hanno* and it has a more prominent white, postdiscal band on the hindwing. From *Leptotes cassius* (above), which also has two spots (the inner one small and faint) in the hindwing tornal area, *C. ammon* may be dis- tinguished by its grey and spotted rather than barred under surface, and by the prominent orange area about the outer tornal black spot. The wing fringes are slightly chequered in *Cyclargus*, but not in *Hemiargus*.

Subspecies

Cyclargus ammon erembis Nabokov, 1948, a subspecies endemic to the Cayman Islands, is more boldly patterned on the underside than is the nominate form from Cuba, and it is a slightly larger and darker insect. It was described from specimens collected by the Oxford University expedition in 1938 and the type locality is West End Point, Little Cayman (Nabokov 1948). *C. a. erembis* was described originally as a full species, and opinion is at present divided as to whether or not this might be its status. Should further research show *erembis* to be specifically distinct, it will become Cayman's first endemic butterfly species.

Female Lucas's Blue, GC (6.ii.2008), RRA

Lucas's Blue at rest on bud of Rabbit Thistle (Tridax procumbens), GC (17.i.2008), RRA

The taxonomy of *Cyclargus* is confusing. *Cyclargus ammon* and *C. thomasi* (Clench, 1941) are treated as separate species by Smith *et al.* (1994) but in Lamas (2004), which we follow here, *thomasi* and its several named forms are all regarded as subspecies of *C. ammon*. *C. a. ammon* (from Cuba) and *C. a. erembis* (Cayman Islands) differ from *C. a. thomasi* and the other subspecies in having a row of three and not four small dark spots at the base of the hindwing underside.

Species' range

Lucas's Blue is found in Florida, the Bahamas, Turks and Caicos Islands, Greater Antilles (discovered in Jamaica in 1985) including the Cayman Islands and Mona Island, Virgin Islands and Leeward Islands south-eastwards to St Kitts.

Cayman Islands distribution

Grand Cayman, Little Cayman, Cayman Brac

Habitat

Like most other Blues, *C. ammon* is a butterfly of sunny, open situations, and in the Cayman Islands it is especially numerous on the beach ridges. Smith *et al.* (1974) remark that it has a preference for rather dry habitats.

History

Carpenter & Lewis (1943) found Lucas's Blue to be abundant on Grand Cayman in 1938, and they also recorded an insect under the name '*Hemiargus catalina* (Fabricius)' on Little Cayman and Cayman Brac which they compared to *C. ammon*. It would appear that '*Hemiargus catalina*' refers to *C. ammon* and not to *Leptotes cassius catilina* (above). In 1975 (Askew 1980) and 2008 *C. ammon* was plentiful on all three Cayman Islands.

Biology

The larva of *C. ammon* has not been found in the Cayman Islands, but adult butterflies are nearly always on or near *Caesalpinia*. *C. bonduc* (Cockspur) is one probable larval food-plant; another possible food-plant is *Acacia farnesiana* (Sweet Acacia), a shrub naturalized in the Cayman Islands. On Cuba the larva is known to feed on *Caesalpinia bahamensis*, *C. pauciflora* and *C. vesicaria*, and also on *Acacia*, *Mimosa* and other Fabaceae. Butterflies take nectar from a variety of herbaceous plants in the Cayman Islands, including *Bidens alba*, *Lippia nodiflora*, *Sesuvium portulacastrum*, *Spilanthes urens* and *Tridax procumbens*.

White Butterflies and Sulphurs

Pieridae

Pieridae is an easily recognizable family, the butterflies being mainly white or yellow to orange with usually simple brown or black markings. The white and yellow colours are due to pterin pigments, derived from metabolic 'waste' products deposited in the wing scales, and among butterflies these are unique to this family (Feltwell & Rothschild 1974).

Two subfamilies of Pieridae are found in the Cayman Islands, Pierinae (Whites, represented by *Glutophrissa* (= *Appias*) and *Ascia*) and Coliadinae (Sulphurs, represented by *Eurema*, *Pyrisitia*, *Abaeis*, *Nathalis*, *Anteos*, *Phoebis* and *Aphrissa*). Most White Butterfly caterpillars feed on cruciferous plants (Brassicaceae), and mustard oils obtained from these foodplants may render them, and perhaps the adult butterflies, distasteful to would-be predators. *Ascia monuste* is rejected as a food item by caged birds (Jacamars) (DeVries 1987). Sulphurs, in contrast, feed as larvae mostly on Fabaceae. In both subfamilies the larvae are elongated and cylindrical, smooth or with a coat of generally very short hairs, and in Coliadinae at least, usually cryptic yellowish to green, often with longitudinal stripes. The pupae also are often cryptic and held against the substrate by a silken girdle.

Adult Pieridae, like Papilionidae and Hesperiidae, have six functional walking legs in both sexes. Sexual dimorphism in wing pattern, though not appearing to be extreme, is usual in Pieridae. Some dimorphism is evident in ordinary light, but is remarkably increased if the butterflies are viewed under ultra-violet light when species and sex specific ultra-violet reflectance patterns are revealed (Silberglied 1984). These are important in mating, being partner recognition signals and, in males, a deterrent to rivals.

Many species of Pieridae are strongly migratory so that island populations tend not to remain isolated for very long. In consequence, subspecific evolution is impeded. As witness to the migratory propensity in the family, a relatively high proportion of Pieridae are occasional vagrants to the Cayman Islands.

Florida White

Glutophrissa drusilla (Cramer, 1777)
Plate IV (1-4)

Recognition
FWL 22-28 mm. The male *Glutophrissa drusilla* is shining white on the upper surface, the hindwing underside is slightly yellowish, and there is an extremely fine, black margin to the rather pointed forewing apex. The female is more creamy white, becoming somewhat transparent with age, and the forewing has a broader dark brown tip which extends down the outer margin, and brown scaling at the base which extends along the costal margin. There is a very thin, short, transverse line at the end of the female's forewing cell. Beneath, the female is cream coloured with a distinct yellow area on the forewing at the base of the cell. There is some seasonal variation in the Florida White, wet (summer) season butterflies

Female Florida White feeding at Jack-in-the-Bush (Chromolaena odorata), CB (29.i.2008), RRA

This is a widespread species ranging from Florida, through Central America and the West Indies, to South America as far south as Brazil.

Cayman Islands distribution
Recorded from all three Cayman Islands, but of regular occurrence only on Little Cayman and Cayman Brac.

Habitat
Glutophrissa drusilla is essentially a shade-loving woodland butterfly. On a transect of Little Cayman in August 1975, when it was the most numerous butterfly species, 135 specimens were seen, 61 percent in ironshore woodland, 31 percent in woodland on bluff limestone, seven percent on the beach ridge and one percent in mangroves (Askew 1980).

History
The Florida White was found on Little Cayman and Cayman Brac in 1938 (Carpenter & Lewis 1943), 1975 (Askew 1980) and 2008, and a few specimens were collected on Cayman Brac in 1990 (Miller & Steinhauser 1992). Two specimens were recorded from Grand Cayman in 1975 (Askew 1980) and Peter Davey tells us that he saw a few flying deep in the forest near the Old Man Bay caves, Grand Cayman, in July 2002. Why *G. drusilla* should be so scarce on Grand Cayman, yet plentiful in the Sister Islands, is a mystery.

Biology
In Central and South America the nominate subspecies of *G. drusilla* makes mass migratory movements, but these have not been observed in *G. d. poeyi*. Carpenter & Lewis (1943) record that in Little

having larger dark areas. The club of the antenna is conspicuously blue-white.

The Florida White is quite like *Ascia monuste* (below), the other frequent white butterfly in the Cayman Islands, but distinguishable by usually being slightly smaller with more pointed forewings, having a different black pattern on the forewing outer margin and a yellow mark on the under surface at the base of the forewing. The Florida White flies more rapidly than the Great Southern White, and is much more likely to be seen in the Sister Islands than on Grand Cayman.

Subspecies
Glutophrissa drusilla poeyi (Butler, 1872) inhabits the Bahamas, Cuba and the Isle of Pines, and the Cayman Islands. In much of the literature it appears under the generic name *Aphrissa*.

Male Florida White, CB (28.i.2008), RRA

Cayman and Cayman Brac, contrary to our own observations, *G. drusilla* was 'a weak flier and was never seen to rise many feet above the ground; flights were short. Weakness of flight probably accounts for the large percentage of lizard-marked wings'. *G. drusilla* has a strong, rather erratic flight, and Smith *et al.* (1994) interpret Carpenter and Lewis's observations of weak flight as indicating that the butterflies had recently arrived after a long overseas migration, but the abundance of strongly flying Florida Whites in Little Cayman in 1975 (Askew 1980) and again in 2008 is consistent with the existence of a permanent breeding population.

Adult Florida Whites nectar at *Chromolaena odorata*. Caterpillars have not been found in the Cayman Islands, but a native vine-like shrub, *Capparis flexuosa* (Raw Bones-Bloody Head) (Capparaceae), is among the reported larval foodplants. The chalcid wasp *Pteromalus* is a gregarious internal parasitoid of pupae of *G. drusilla*.

Great Southern White

Ascia monuste (Linnaeus, 1764)
Plate IV (5-7)

Recognition
FWL 22-32 mm. (dwarf specimens quite frequent). The upper surface of the male *Ascia monuste* is white except for some triangular, black marginal markings at the ends of the veins on the forewing. The size of these markings varies between individuals, but they are largest at the wing apex and diminish posteriorly. In the female the ground colour is cream to slightly dusky, and the black marginal markings are larger and confluent, the series continuing onto the ends of the hindwing veins. The underside of the hindwing and apex of the forewing, in both sexes, are cream coloured to slightly brownish, and the female hindwing under surface has some of the veins a dusky brown with similarly coloured, vague and variable dusky markings between the veins. The club of the antenna is a conspicuous whitish blue as in the Florida White.

Male Great Southern White, GC (9.ii.2006), MLA

Male Great Southern White, GC (20.x.2007), KDG

Subspecies
Ascia monuste eubotea (Godart, 1819) (= *A. m. evonima* (Boisduval, 1836)) occurs throughout the Greater Antilles and associated islands.

Species' range
The Great Southern White is distributed from the south of the United States, through Central America and the Caribbean islands, into South America as far south as Patagonia.

Cayman Islands distribution
A butterfly found on all three Cayman Islands with the highest density probably on Grand Cayman, although the situation may fluctuate and could be quickly altered by immigrating swarms (but see below under Biology).

Habitat
On Grand Cayman, *Ascia monuste* may occur anywhere, but it is particularly plentiful along the edges of roads, flying along hedges, and in gardens, parks and agricultural land. In a transect of Little Cayman in 1975, when the species was not very common, only eleven specimens were counted (compared to 135 *Glutophrissa drusilla*) and most of those were on the coast (Askew 1980).

History
Ascia monuste was found on all three Cayman Islands in 1938 and recorded under the name *Pieris phileta phileta* Fabricius (Carpenter & Lewis 1943). It was least abundant on Little Cayman, but found in swarms on Cayman Brac.

In 1975 (Askew 1980) and 2008 it was again found on all three Cayman Islands (Askew 1980), its status usually being common to abundant.

Biology

The Great Southern White sometimes appears in very large swarms. Migratory movements have been observed frequently in Central and South America, but mass flights between islands have not been reported in the West Indies (Smith *et al.* 1994). The swarms of butterflies that sometimes occur on Grand Cayman are therefore more likely to be locally bred than immigrants. Carpenter & Lewis (1943) reached a similar conclusion, and Nielsen (1961) in Florida showed that only the darker *phileta* form of *A. monuste* is migratory, paler forms (such as occur in the Cayman Islands) remaining resident.

Female Great Southern White, GC (31.i.2006), MLA

Female Great Southern White, GC (9.ii.2006), MLA

In January 2006 the butterflies were so dense along the south coast approaching George Town that huge numbers were killed by traffic, and their bodies littered the roadsides like drifting snow.

Adult butterflies have been seen nectaring at *Croton linearis, Jatropha integerrima* and *Stachytarpheta jamaicensis*. They are unpalatable to at least some insectivorous birds (DeVries 1987).

On Cayman Brac in 1940, C. Bernard Lewis observed butterflies congregating in open areas, about an hour before sunset, before settling down for the night. They often roosted in bushes of what was stated to be *Croton linearis* (Rosemary) (Euphorbiaceae) on which many of their caterpillars were feeding (Carpenter & Lewis 1943), but *C. linearis* was probably misidentified as it is a very unlikely foodplant. On Grand Cayman, eggs and larvae of *A. monuste* have been found on *Cakile lanceolata* (Coastal Sea Rocket) (Brassicaceae) and *Capparis flexuosa* (Raw Bones-Bloody Head) (Capparaceae). Larvae are known to eat a range of other plants including *Cleome* (Capparaceae), *Batis* (Bataceae), *Lepidium* (Brassicaceae)

*Larva and pupa of Great Southern White on Bloody Head-Raw Bones (*Capparis flexuosa*), GC (21.vii.2002), AS*

and cultivated brassicas such as cabbage and kale. It can be a horticultural pest. The yellow spindle-shaped eggs are laid in clusters. The fully grown larva are also gregarious, green above, brownish ventrally, with a yellow mid-dorsal line, purple-grey longitudinal stripes, purplish spots, and a covering of pale hairs.

Barred Sulphur

Eurema daira (Godart, 1819)
Plate V (1,2)

Recognition
FWL 15-17 mm. This is a small butterfly with rather rounded wings. The sexes are dissimilar. In the male, the forewing upper surface is yellow with a broadly black apex and a weakly curved grey bar, its posterior edge lined with orange, running parallel and close to the inner wing margin; the hindwing upper surface is white with a black outer margin. The female is white above with dark borders to both wings.

The under surface of both sexes is shining white to pale buff with the dark areas of the upper surface showing through, and there is sometimes faint, brown stippling on the hindwing.

There is some seasonal variation in the subspecies of *Eurema daira* which flies in the Cayman Islands, the so-called wet season form having more extensive dark markings than the dry season form, and a more shining white, rather than buff, under surface.

Subspecies
Eurema daira palmira (Poey, 1852) inhabits the Greater and Lesser Antilles. The nominate subspecies, common in Florida and elsewhere in the south-eastern United States and found in the Bahamas and occasionally Cuba (Smith *et al.* 1994), is usually yellow on the upper surfaces of both wings in both sexes.

Species' range
Eurema daira has a broad range encompassing the southern United States, Central America, the Bahamas, West Indies and South America as far south as Brazil.

Cayman Islands distribution
Grand Cayman, Little Cayman, Cayman Brac

Habitat
This is a butterfly of short vegetation, flying in weedy, uncultivated places, roadsides and track edges, and open scrubland.

History
The 1938 Oxford University expedition to the Cayman Islands failed to find the Barred Sulphur. It was discovered first in

1975 when it was uncommon and local on both Cayman Brac and Little Cayman (Askew 1980). Schwartz *et al.* (1987) added it to the Grand Cayman faunal list, finding butterflies in 1985 at George Town, Cayman Kai, Old Man Bay and Boatswain Bay. Miller & Steinhauser (1992) found *E. daira* on Grand Cayman (Seven Mile Beach and Great Beach) and Cayman Brac (Brac Reef Resort, Jennifer Bay and Hawkesbill Bay) in October and November 1990. A specimen in the National Trust was captured in George Town on 5 December 2000. The Barred Sulphur was not found in any of the Cayman Islands in 2008 despite close scrutiny of dozens of butterflies which all turned out to be the following species.

Biology

Eurema daira, like other species of *Eurema* and *Pyrisitia* in the Cayman Islands, is not a fast-flying butterfly, but it is nonetheless elusive as it flits very low amongst stems and leaves, and nectars at herbaceous flowers. *Bidens alba* has been noted as a nectar plant, and the butterflies sometimes imbibe liquid from damp soil or sand (Smith *et al.* 1974) or cattle dung (DeVries 1987). The mainly green larvae are known to feed on a range of Fabaceae, but have not been observed in the Cayman Islands.

In Costa Rica, where *E. daira* is a very common butterfly, dry and wet season forms are distinct, not only in appearance but also in behaviour. The wet season form is active and dispersive, whilst the dry season form is relatively sedentary, is in reproductive diapause and aggregates in shady places during much of the day (DeVries 1987).

False Barred Sulphur

Eurema elathea (Cramer, 1777)
Plate V (3-7)

Recognition

FWL 14-17 mm. *Eurema elathea* is very like the preceding species, and the sexes and seasonal forms differ much as in *E. daira*. The male has a dark bar on the upper surface of the forewing parallel to the inner margin. As in *E. daira* this bar is lined posteriorly by orange, but it is straight in *E. elathea,* not weakly curved, and black rather than grey. In the dry season form this bar can be very much reduced or even absent, as in a specimen in the National Trust reference collection found at East End, Grand Cayman in January 2005. Female *E. elathea* in the Cayman Islands differ from female *E. daira* in always having at least some yellow scaling on the forewing upper surface. There is some brown stippling on the buff under surface of the hindwing in both sexes but especially the female. Brown markings are most pronounced in the so-called wet season form, but since both forms can be seen flying together, the implication that the variation is seasonal requires further investigation.

Subspecies

There are no described subspecies. Across its range this is a relatively uniform species, in contrast to the allied *E. daira*.

Species' range

Eurema elathea is distributed from Central America to Brazil, and in the Bahamas, and Greater and Lesser Antilles.

Cayman Islands distribution

Grand Cayman, Little Cayman, Cayman Brac

Male False Barred Sulphur, GC (31.i.2006), RRA

Female False Barred Sulphur, CB (26.i.2008), RRA

Two males of False Barred Sulphur courting a female but attempting to mate with each other, GC (29.i.2006), RRA

Pair of False Barred Sulphur in copulation, GC (4.ii.2006), RRA

Habitat

This is another butterfly of short vegetation, flying in open, grassy places such as road verges, broad track edges and in sparse scrubland.

History

In contrast to *E. daira*, *E. elathea* was recorded on Grand Cayman by Carpenter & Lewis (1943) and has remained abundant on the island ever since, easily the most numerous species of *Eurema*. It was not found on Cayman Brac prior to the capture, late in 1985, of four specimens at the airport (Schwartz *et al.* (1987), but in 2008 *E. elathea* was abundant on both Cayman Brac and Little Cayman, being noted for the first time on the latter island.

Biology

The greenish caterpillar of *E. elathea* has been reared on *Stylosanthes hamata* (Donkey Weed) (Fabaceae) in Jamaica

(Brown & Heineman 1972). This plant grows on Grand Cayman and we believe it to be the larval food-plant of both *E. elathea* and *E. daira* when allowed to grow to a height of twenty centimetres or so.

The butterfly flies low and rather erratically, the males for quite long periods, perching infrequently with wings closed. Nectaring has been observed on *Spilanthes urens*.

Shy Sulphur

Pyrisitia messalina (Fabricius, 1787)
Plate V (8-10)

Recognition
FWL 10-18 mm. (the wide variation in size is remarked upon by Carpenter & Lewis (1943)). Species of *Pyrisitia* resemble those of *Eurema*, in which genus they used to be placed, but males lack the dark bar subparallel to the forewing inner margin and in females there is a purplish brown to rust-coloured apical marginal spot on the hindwing under surface. Forewings of *Pyrisitia messalina* are more rounded than in the two preceding species of *Eurema*. The male upper surface is white, without a dark bar, but with a continuous black border to the outer margins of both fore- and hindwings, broader on the former. The female also has a white upper surface, but the black borders are reduced to apical patches on both wings and small points at the ends of the veins. The under surface of the forewing is white with a broad, yellow costal margin and a small black subapical spot, larger in the female. The hindwing under surface is pale yellow with a scattering of dark specks and an apical purplish angular mark with a black spot at its inner apex;

these markings are again larger and more distinct in the female.

Subspecies
Caymanian insects do not differ from the nominate form described from Jamaica.

Species' range
This is a butterfly with a limited range, occurring in the Bahamas (possibly as a distinct subspecies), Cuba and the Isle of Pines, Grand Cayman (not reported after 1938) and Jamaica. It may at one time have occurred in Florida (Smith *et al.* 1994).

Cayman Islands distribution
Grand Cayman (extinct?)

Habitat
Lewis in Carpenter & Lewis (1943) writes of the Shy Sulphur on Grand Cayman 'This butterfly, like [*Eurema*] *elathea*, flies close to the ground, but remains even more in the bushy (but not forest) areas, rarely coming into the open.' Similarly in the Bahamas and Jamaica, Smith *et*

Four mounted specimens of Shy Sulphur in the Natural History Museum, London, GC (v & vi.1938) collected by C.B. Lewis and G.H. Thompson at 'west end of Georgetown'. Males upper and under surfaces above, females the same below.

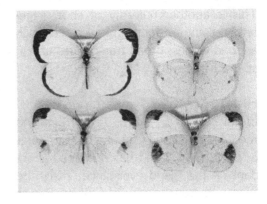

al. (1994) observe that it is a species that prefers the shade and protection of the bush and shady wood edges.

History

The 1938 Oxford University expedition collected a sample of seven males and twelve females on Grand Cayman (Carpenter & Lewis 1943). Of these, four males and nine females have been examined in the Natural History Museum (London), all labelled as being collected at the 'west end of Georgetown' on various dates between 20 April and 14 June 1938. Although searched for, *P. messalina* has not been rediscovered in the Cayman Islands.

Biology

The larvae feed on a range of species of *Senna* (Fabaceae) in Jamaica and on *Desmodium* in Cuba (Smith *et al.* 1994).

Little Sulphur

Pyrisitia lisa (Boisduval & Leconte, 1830)
Plate V (11-14)

Recognition

FWL16-17 mm. The ground colour of all wing surfaces of both sexes is yellow but paler, occasionally cream-coloured, in females, and the upper surfaces have complete (male) or incomplete (female) black borders extending down from the very broad, black forewing apex. The forewing is relatively pointed and the black wing-tip has a straight inner edge. There is a small, black point at the end of the forewing cell. The hindwing underside has small, brown stipplings and a reddish subapical spot, best developed in the female.

Male Little Sulphur (injured), CB (29.i.2008), RRA

Subspecies

Pyrisitia lisa euterpe (Ménétriés, 1832) is the West Indian subspecies, but differences between it and the nominate subspecies, which flies in North America, Bermuda and the Bahamas, are very slight.

Species' range

The Little Sulphur is distributed from the southern United States, Mexico to Costa Rica, Bermuda, the Bahamas, Greater and Lesser Antilles east to Barbados.

Cayman Islands distribution

Grand Cayman, Little Cayman, Cayman Brac

Habitat

Like most other species of *Eurema* and *Pyrisitia*, this is a low-flying butterfly of open areas with short vegetation.

History

Pyrisitia lisa was found in 1938 on Grand Cayman by the Oxford University expedition, and described as very common in all open parts, but it was not seen on Little Cayman or Cayman Brac. On Grand Cayman it remained abundant in 1975 and 1985, and was recorded from three localities in 1990 (Miller & Steinhauser 1992). It was quite plentiful on both Cayman Brac and Little Cayman in 1975 (Askew 1994), but only a few individuals were seen in the Sister Islands in 2008. It was rare on Grand Cayman in 1995, 1997 and 2002, and has not been seen there since.

Biology

The immature stages of *P. lisa* have not been discovered in the Cayman Islands, but in Jamaica larvae are known to feed on Fabaceae, several species of *Senna*, *Desmanthus virgatus* (Ground Tamarind) and *Mimosa pudica* (Shame-face) being recorded (Smith *et al.* 1994). In Cuba eggs are laid on the upper side of the mid-vein of a leaf of *M. pudica,* and the ovipositing butterfly is not disturbed by the closing of the leaflets against her abdomen (Dethier 1940). The fully grown caterpillar is green with darker green mid-dorsal and dorso-lateral longitudinal stripes, a yellow stripe on each side, and white spiracles.

Large migratory swarms of the North American nominate subspecies *P. l. lisa* have been reported off Bermuda (Brown & Heineman 1972), but *P. l. euterpe* shows little tendency to migrate. It is not a strong flier, and takes little evasive action when disturbed.

Mimosa Sulphur, Blacktip Sulphur

Pyrisitia nise (Cramer, 1775)
Plate V (15,16)

Recognition

FWL 13-15 mm. This is a small, yellow butterfly with rather rounded wings. The sexes are similar, although the male is more deeply coloured than the female. The forewing has a moderately broad black apex which tapers down the outer margin to reach, or not quite reach, the anal angle, but there is only a tiny, scarcely visible, black spot at the end of the cell. The hindwing border is very narrow, usually reduced to small points at the ends of the veins, and there is a little faint, brown scrawling on the under surface which, in the female, bears the reddish apical spot displayed by species of *Pyrisitia*.

Subspecies

Pyrisitia nise nise, the nominate subspecies, is the form found in the West Indies.

Species' range

The distribution of *P. nise* is from Texas and southern Florida (irregular occurrence) through Central and South America to Argentina, and in the Bahamas, Cuba, Isle of Pines and Jamaica. It is apparently a very occasional vagrant to the Cayman Islands.

Cayman Islands distribution

Cayman Brac (one record)

Habitat

Uncultivated ground with secondary vegetation, often at woodland edges or in light scrub.

History

The capture of a single male on 5 November 1990 on Cayman Brac, on the South Side road between Jennifer and Pollard Bays, in secondary scrub and trees below a cliff, is reported by Miller & Steinhauser (1992). In the abstract to their paper, the authors refer to the insect as *Eurema dina dina* (Poey, 1832), but this is clearly an error, confirmed by the photograph in their paper. The specimen is small, the forewing length being only 13 mm as indicated by the scale-line on the photograph. This is the only record of *P. nise* in the Cayman Islands.

Biology

The larva is known to feed on *Desmanthus virgatus*, *Mimosa pudica* and species of *Senna* in Jamaica (Smith *et al.* 1994), a similar range of food-plants to that of *P. lisa* (above). When fully grown, the caterpillar is bluish green with whitish lateral stripes and white spiracles.

The butterfly keeps close to the ground, not straying far from cover, and its flight, when threatened, is agile and evasive, contrasting with the behaviour of *P. lisa* for which it could be mistaken (Brown & Heineman 1972).

Black-bordered Orange, Sleepy Orange

Abaeis nicippe (Cramer, 1779)
Plate V (17-19)

Recognition

FWL 22-25 mm. This species is easily recognized, being distinguished from the species of *Eurema* and *Pyrisitia* that fly in the Cayman Islands by its orange (rather than yellow or white) upper surface with broad, black borders (complete in the male, almost complete in the female) and larger size. There is a small, transverse dark line at the end of the forewing cell on both surfaces in both sexes. In the male, the black margin to the hindwing is expanded in the middle to produce an inwardly directed tooth, and the orange ground colour is brighter than in the female.

Subspecies

None described

Species' range

Abaeis nicippe is found in the United States (mainly in the south-east but occasionally spreading north as far as New York and New England), Central America south to Costa Rica (very rare), the Bahamas, Greater Antilles and associated islands.

Cayman Islands distribution

Grand Cayman, Little Cayman, Cayman Brac

Habitat

This is a sun-loving insect which flies low and erratically over open, grassy areas.

History

Abaeis nicippe has never been a common insect in the Cayman Islands. It was recorded from the George Town area of Grand Cayman, and from the north-east of Cayman Brac, by Carpenter & Lewis (1943), and in 1975 it was found on Little Cayman as well as on Grand Cayman and Cayman Brac (Askew 1980). Since 1975, however, sightings of *A. nicippe* have been sporadic. A single specimen was collected in the Great Beach area of Grand Cayman in November 1990 (Miller & Steinhauser 1992), and a female found by Peter Davey on Cayman Brac in February 1994 is in the National Trust reference collection. A small colony of *A. nicippe* was observed on a patch of waste ground in south George Town, Grand Cayman, during August and September 2001, but unfortunately it was destroyed by an application of weed-killer.

Biology

Eggs are laid singly on leaves of *Senna* (Fabaceae). The larva is green speckled with black and has yellow lateral stripes and white spiracles. It is covered in short, white hairs. The pupa has two intergrading colour forms, one greenish more or less speckled and blotched with dark brown, and the other almost entirely dark brown (Scudder 1889).

The Black-bordered Orange is more an insect of temperate than tropical conditions. In the United States, it extends northwards and westwards annually in summer beyond its permanent breeding range in the south-east. In the Caribbean it seems to be of regular occurrence only in some of the Bahamas, and in Cuba and Hispaniola, elsewhere appearing sporadically and erratically. It is not found south of Costa Rica or east of Puerto Rico.

Dainty Sulphur

Nathalis iole Boisduval, 1836
Plate V (20,21)

Recognition

FWL 12-14 mm. *Nathalis iole* somewhat resembles a small *Eurema,* but has narrower wings and more extensive dark markings. On the upper surface the ground colour is yellow on the forewing

The two recorded Caymanian specimens of Dainty Sulphur, male (above) from GC (22.viii.1985) and female from LC (2.viii.1975).

and orange-yellow on the hindwing, the contrast much more apparent in females. The forewing has a broad, dark brown apex and a dusky bar along the inner margin which is expanded to a wedge-shape in the female. The hindwing has a blackish costal margin, interrupted basally in the male by a small yellowish androconial patch, and blackened ends to the veins, the latter dark markings larger in the female which has an additional submarginal transverse dark band. On the posterior half of the under surface of the forewing in both sexes there are two submarginal black marks, the anterior of which is round, half the size of the posterior, and visible also on the upper surface.

Subspecies
None recognized

Species' range
A butterfly distributed from the Great Lakes in the United States through Central

America south to Colombia, and in the Bahamas and northern Greater Antilles (Cuba, Jamaica, Hispaniola). It has been very occasionally found in the Cayman Islands.

Cayman Islands distribution
Grand Cayman, Little Cayman (one specimen reported from each island)

Habitat
Nathalis iole flies over dry open ground with mostly short vegetation.

History
The first specimen found in the Cayman Islands is a female caught on Little Cayman on 2 August 1975. It was flying half a mile north of Blossom Village in dry, open, sun-baked terrain with *Evolvulus arbuscula*, a place inhabited by the Little Cayman endemic snail *Cerion nanus* (Maynard) (Askew 1980). The second, and only other known Caymanian specimen, is a male from Grand Cayman, found on 22 August 1985 in the Great Beach area in dry pastureland. Both of these insects were in fresh condition when caught, indicating that they had probably only just emerged as adults having developed close to their places of capture. The paucity of Cayman Islands records strongly suggests, however, that *N. iole* is a temporary colonist, not a permanent resident.

Biology
The larva of *N. iole* is reported to feed upon a variety of Asteraceae, but *Bidens* appears to be the most widely used foodplant (Smith *et al.* 1994). In Cuba the fully grown caterpillar varies from grass green to dark green, with a purple middorsal line and yellowish lateral lines. The smooth, green pupa is unusual in Pieridae in lacking a frontal process.

Adult insects fly close to the ground, like species of *Eurema* and *Pyrisitia*, and they appear early in the morning. In Jamaica and the United States *N. iole* tends to be more numerous at higher altitudes, and Brown & Heineman (1972) draw attention to its cold hardiness, it being 'among the last of the non-hibernating butterflies to disappear in the early winter, the end of November, on the high plains of Colorado'. It is interesting that the Dainty Sulphur has been found in very hot places in the Cayman Islands; the butterfly is evidently tolerant of a wide range of temperatures.

Giant Brimstone, Yellow Angled Sulphur

Anteos maerula (Fabricius, 1775)
Plate IV (8)

Recognition
FWL 39-46 mm. *Anteos maerula* is the largest pierid in the Cayman Islands and its size, distinctive wing shape and yellow colour easily distinguish it. The male upper surface is clear yellow with a conspicuous dark spot at the end of the forewing cell, and a smaller, fainter spot at the end of the hindwing cell. The female is similar but usually with a less bright, pale greenish ground colour. Both sexes have characteristically shaped wings, the forewing falcate and the hindwing with a projecting angle in the middle of the outer margin.

Subspecies
The separation of West Indian populations as a distinct subspecies is unjustified (Smith *et al.* 1994).

Species' range

A species which ranges from southern Florida, where it does not breed (Gerberg & Arnett 1989), through Central America to Colombia and Peru, in the Greater Antilles, and in St Kitts and Guadeloupe in the Lesser Antilles.

Cayman Islands distribution

Grand Cayman (vagrant)

Habitat

This strong-flying butterfly is found in scrubland and open ground, and also in gardens. In the Dominican Republic it occurs from sea-level to well over 2,000 m. (Smith *et al.* 1994).

History

C. B. Williams (1930) recorded a north-westerly migration of a few *A. maerula*, among large numbers of *Phoebis sennae*, flying over the sea off the Cayman Islands (Brown & Heineman 1972). On Grand Cayman it was first observed on 21 August 1985 (Askew 1988) when a solitary male was taken flying over scrubby pastureland at Botabano. Another was seen on 2 October 1995 in a garden at Cayman Kai, and sightings in March 2002 of high flying yellow butterflies at North Sound Estates and the south end of the Mastic Trail were probably of *A. maerula*.

Biology

In the Greater Antilles, larvae of *A. maerula* have been found feeding on several different species of *Senna* (Fabaceae), particularly *S. emarginata*, and *Gliricidia sepium* (Fabaceae) may be another foodplant. The fully grown caterpillar is green with white lateral lines and, on the dorsal surface, black transverse bars edged with turquoise (Smith *et al.* 1994).

The butterflies soar 'to great heights with a powerful, majestic flight' (Brown & Heineman 1972), descending to visit flowers of *Bougainvillea, Caesalpinia, Cordia, Hibiscus* and other flowering shrubs. They seem to have a preference for red flowers.

Cloudless Sulphur

Phoebis sennae (Linnaeus, 1758)
Plate IV (9,10)

Recognition

FWL 28-34 mm., the smaller individuals often being females. Of the four medium-sized to large yellow butterfly species recognized in the Cayman Islands, *Anteos* (above) and three species of *Phoebis*, *Phoebis sennae* is the most numerous and widespread. The male upper surface is clear yellow, unmarked, with androconial scales forming slightly paler marginal bands, broadest on the forewing, visible when the wings are viewed at an oblique angle. The female upper surface ground colour varies from cream to pale pinkish yellow (peach colour), slightly deeper on the hindwing. The forewing has a narrow brown outer marginal border composed of almost contiguous small marks at the ends of the veins, some faint, brown markings near the apex, and a prominent, rounded, pale-centred brown spot at the end of the forewing cell. The hindwing has small brown marks at the ends of the veins, but is otherwise unmarked although the underside pattern may show through faintly. The pattern of dark markings on the under surface is similar in both sexes, but the spots and stipples are larger in the female.

Female Cloudless Sulphur, CB (27.i.2008), RRA

Females of the Cloudless Sulphur resemble those of *Phoebis agarithe* (below), but can always be identified by the purplish brown postdiscal spot-line on the forewing under surface, subparallel to the outer wing margin, being dislocated and sharply angled inwards at vein 4 (at the level of the cell end spot). In *P. agarithe* this line is straight. In addition, the spot in the forewing cell is much larger in *P. sennae* than in *P. agarithe* and the forewing outer marginal border of brown spots is better defined, but the forewing apex does not have a distinct brown tip in *P. sennae* as it does in *P. agarithe*.

Subspecies

West Indian specimens are usually attributed to *P. sennae sennae*, but other described subspecies are scarcely distinguishable.

Species' range

The Cloudless Sulphur occurs from Florida and the Gulf States, through Central America, south to Uruguay, and in the Bahamas and throughout the West Indies. It is probably the commonest and most widespread species of Pieridae in the New World.

Cayman Islands distribution

Grand Cayman, Little Cayman, Cayman Brac

Habitat

A butterfly which may be seen almost anywhere except in mangroves and dense woodland, equally common in parks, gardens, along hedgerows and in more open areas.

History
The 1938 Oxford University expedition found *P. sennae* 'a common species in open parts of each island' (Carpenter & Lewis 1943). Its status has not since changed.

Biology
Phoebis sennae lays its eggs singly on buds and fresh foliage of several species of *Senna* (Fabaceae). On Grand Cayman, *Senna surattensis* (Scrambled Eggs Cassia), *Senna alata* (Candle Bush) and *S. occidentalis* (Dandelion, Septic Weed) have been confirmed as food-plants. The fully grown caterpillar is yellowish with transverse rows of small black points, orange-yellow lateral lines and blue-black intersegmental, dorsal, transverse lines. It feeds mainly at night, living during daytime in a tent of spun leaves. The pupa is very dumpy, usually green but occurring also in a pink form.

The butterfly is strong-flying and well known to undertake migrations involving enormous numbers of individuals. Brown

Cloudless Sulphur pupa, GC (9.ii.2003), AS

and Heineman (1972) mention a paper by J. W. Plaxton, published in 1891, in which is described 'a tremendous migration [of *P. sennae*] during May and June at Kingston [Jamaica] and at sea off the Caymans, with the direction of flight towards the northwest'. Adult *P. sennae* feed especially on flowering shrubs and taller herbaceous plants; in the Cayman Islands *Asystasia gangetica*, *Caesalpinia pulcherrima*, *Colubrina cubensis*, *Cordia* species, *Duranta erecta* and *Hibiscus* have been confirmed as nectar plants. In other parts of its range, male *P. sennae* have been seen congregating and drinking at mud puddles.

In *Phoebis* ultraviolet wing patterns (*i.e.* patterns that become visible to us only when ultraviolet wavelengths are shone on the butterflies, but are visible to butterfly eyes in normal light) are species specific and believed to be important in inter- and intraspecific behaviours (Silberglied 1977).

Cloudless Orange, Large Orange Sulphur

Phoebis agarithe (Boisduval, 1836)
Plate IV (11)

Recognition
FWL 31-35 mm. The male upper surface is orange, more yellow on the hindwing than on the forewing. The female ground colour varies between orange-yellow and cream, and the upper surface of the forewing has a brown tip (a larger mark than in *P. sennae* (above)), brown marginal marks at the ends of the veins and a brown cell-end spot (these latter two smaller than in *P. sennae*). The forewing under surface in both sexes has a straight, unbroken post-discal line, fainter in the male, which reli-

ably distinguishes *P. agarithe* from *P. sennae*. Additional characters for separating females of these two species are given under *P. sennae* (above).

Phoebis argante (Fabricius, 1775), another large, orange pierid which has not yet been reported from the Cayman Islands but which might turn up here, has the forewing under surface postdiscal line broken at vein 4 as in *P. sennae*, but with the two halves of the line less strongly offset.

Subspecies
Phoebis agarithe antillia Brown, 1929 is the West Indian subspecies; other subspecies are found on the mainland.

Species' range
The Cloudless Orange is another widely distributed butterfly, found from the southern United States, through Central America south to Peru, in the Bahamas and almost throughout the West Indies.

Cayman Islands distribution
Grand Cayman, Cayman Brac

Habitat
Phoebis agarithe is most frequently seen in parks and gardens in the Cayman Islands. It has a greater preference than other species of *Phoebis* for drier situations (Brown & Heineman 1972).

History
Phoebis agarithe has a somewhat chequered history in the Cayman Islands. It was not seen in 1938, although Carpenter & Lewis (1943) state that it had been found on Grand Cayman at an earlier unspecified date, and it was not observed in 1975 either. A specimen was taken at Boatswain Point on Grand Cayman on 23 September 1983 (E. J. Gerberg, personal communication). In August 1985 it was common in George Town and other places in the west of Grand Cayman (Askew 1988) and Schwartz et al. (1987) report the collection at the end of November 1985 of four specimens at Boatswain Bay, Grand Cayman, and a single specimen from Cayman Brac airport, the first record for the Sister Island. It was found sparingly on Cayman Brac in January 2008. *P. agarithe* has remained fairly frequent in and around George Town and the western part of Grand Cayman since the mid-1980s at which time it probably recolonized the island.

Biology
The immature stages of *P. agarithe* have not been observed in the Cayman Islands, but the larvae are reported to feed upon *Cassia* and *Senna* species, *Pithecellobium unguis-cati* (Privet, Catclaw Blackbead) (Fabaceae), and other leguminous shrubs (DeVries 1987, Gerberg & Arnett 1989, Smith *et al.* 1994). In George Town gardens the butterflies are often seen flying about *P. unguis-cati*.

Phoebis agarithe and *P. sennae* often fly together and both have a swift, but not necessarily high, flight. Females are more easily caught than males. Both sexes visit flowers of *Bougainvillea*, *Hibiscus*, *Lantana*, *Stachytarpheta* and other flowering plants.

Orange-barred Sulphur
Phoebis philea (Linnaeus, 1763)
Plate IV (12,13)

Recognition
FWL 40-45 mm. The sexes of *Phoebis*

philea differ, but both are basically large, yellow butterflies with conspicuous red-orange areas on the upper surfaces. The male has a broad red-orange bar across the forewing cell, and the hindwing is broadly margined with the same colour. In the female this hindwing marginal band is broader and redder than in the male, but there is only a little red-orange scaling on the forewing which has a dark cell-end spot and marginal and submarginal series of brown spots. The underside is yellow-orange in the male and rusty orange in the female. The yellow and red-orange colouration, coupled with powerful, high flight, is characteristic of *P. philea*.

Subspecies

The subspecies of *P. philea* occurring in the Cayman Islands has not yet been identified; we have been unable to examine any specimen in the hand. It is most likely to be the nominate subspecies, which is the mainland form, but is migratory and has become established in Cuba and Puerto Rico. *P. p. philea* is distinguished by the forewing upper surface lacking a brown cell-end spot in the male, and by being predominantly yellow in the female. *P. p. thalestris* (Illiger, 1801) from Hispaniola has, on the upper surface of the male forewing, a brown cell-end spot and the red-orange bar extends almost to the inner wing margin. The female upper surface is mainly apricot, more heavily marked with brown. *P. p. huebneri* Fruhstorfer, 1907 is a Cuban endemic which resembles *P. p. thalestris* but has a well-defined brown border on the underside of the hindwing.

Species' range

Phoebis philea colonized Florida from Central America in the 1920s (Smith *et al.* 1994), and its distribution extends south to Argentina and Peru, including Grand Bahama, Cuba, Hispaniola and Puerto Rico, with records from Mona Island and Grand Cayman.

Cayman Islands distribution

Grand Cayman (vagrant or recent colonist)

Habitat

Phoebis philea has been seen on Grand Cayman only in large gardens and parks with mature trees. It is reported elsewhere to frequent 'open xeric scrub forest' and 'coffee plantations and other open areas within deciduous forest' (Smith *et al.* 1994).

History

This species is included in the butterfly fauna of the Cayman Islands on the basis of sight records only, but it is so distinctive an insect that we are confident of the identification. Males were first observed in 2002 in Grand Cayman, in south George Town in January and in gardens at West Bay and George Town in March. Thereafter specimens were seen fairly regularly at both of the latter localities until Hurricane Ivan struck in September 2004. After the hurricane, the species was not seen for a long time, but sightings recommenced in February 2007 in a park in George Town. None was observed in 2008.

Biology

Reported food-plants of the larvae of *P. philea* are *Caesalpinia pulcherrima* (Pride of Barbados), *Cassia* and *Senna* species and *Pithecellobium unguis-cati* (Catclaw Blackbead) (Fabaceae).

Males of the butterfly, the only sex seen

on Grand Cayman, fly very fast, usually at a height of five to eight metres. Smith *et al.* (1994) observe that in Haiti it often flies in the forest canopy and is very difficult to capture, descending only very infrequently to feed at blossoms on flowering shrubs such as *Bougainvillea* and *Hibiscus*.

Migrant Sulphur

Aphrissa statira (Cramer, 1777)
Plate IV (14-16)

Recognition
FWL 26-33 mm. Species of *Aphrissa* are yellow and creamy white butterflies, quite similar to *Phoebis* but smaller with rather broader wings and more rounded forewing apex. The male of *A. statira* is yellow above with the outer half (or rather less) of the forewing, and a broad marginal band on the hindwing, creamy white with androconial scales giving these areas a floury or mealy appearance. The female varies from pale yellow to white and has a very narrow, black forewing border on the distal half of the costal margin and anterior three-quarters of the outer margin. There is a small, black spot at the end of the forewing cell in the female. On the under surface the male is pale yellow with the posterior part of the forewing shiny white, the female similar but lightly marked with pale purple-brown postdiscal stipples, forewing margin and forewing cell-end spot.

A. *statira* closely resembles *A. neleis* (Boisduval, 1836), but the latter can always be distinguished in both sexes by the presence, at the very base of the under surface of the hindwing, of a small red spot.

Subspecies
Aphrissa statira cubana d'Almeida, 1939 flies in Cuba, Jamaica and the Cayman Islands. Several other subspecies have been described from elsewhere.

Species' range
The Migrant Sulphur occurs from the southern United States, through Central America and South America to Brazil, and in most of the West Indies.

Cayman Islands distribution
Grand Cayman

Habitat
In the Cayman Islands, A. *statira* has been seen only locally in suburban parks and gardens.

History
A 'fine fresh male' was taken on Grand Cayman by the 1938 Oxford University expedition but was misidentified as 'Phoebis neleis' (Carpenter & Lewis 1943). This record is repeated under the name *Aphrissa neleis* by Schwartz *et al.* (1987).

Pair of Migrant Sulphur in copulation, the male above, GC (26.i.2006), RRA

Dead female Migrant Sulphur caught in spider's web, GC (27.i.2006), RRA

The specimen is in the Natural History Museum, London where its true identity has been ascertained (Askew 1988). The species was again found on Grand Cayman in October 1995 when a female was captured in George Town (Askew), but it was then not seen until 2006 when fresh specimens were most unexpectedly found in abundance in a garden in Bodden Town. On 25 January they were flying with *Ascia monuste* along a hedge, and a number of mating pairs were observed, including a male coupled with a headless female. Females were of both the yellow and white forms. Numbers began to decline from the end of January, and none was seen after 8 February. There was no trace of *A. statira* at the Bodden Town location in January or February 2008.

Biology

The larval food-plant of *A. statira* on Grand Cayman is not known. Elsewhere, caterpillars are reported to feed on a variety of plants: *Callichlamys latifolia* (Bignoniaceae), *Calliandra cubensis* (Powderpuff), *Senna* species, *Dalbergia ecastophyllum* (Coin Vine) and *Entada gigas* (Fabaceae), and *Melicoccus*

bijugatus (Ginep) (Sapindaceae) (Riley 1975, DeVries 1987, Smith *et al.* 1994). DeVries (1987) considers it a distinct possibility that there are two species under the name *Aphrissa statira*, an opinion based on dimorphism of immature stages correlated with different larval food-plant families, one form feeding on legumes (Fabaceae), the other on Bignoniaceae.

The butterfly is strongly migratory (Williams 1930), and when migrating it has been shown able to adjust its direction of flight to compensate for wind drift (Srygley *et al.* 1996). It flies more slowly than a *Phoebis* and is not difficult to catch. Males are most in evidence before mid-morning, patrolling back and forth along hedges, and probably also woodland edges, at a height of one or two metres. In the heat of the day, they are said to fly around the tops of trees (Brown & Heineman 1972). Both sexes are reported to visit a range of red flowers, and males drink at mud puddles (DeVries 1987). On Grand Cayman the butterflies have been seen nectaring at *Lantana* and *Stachytarpheta*.

Orbed Sulphur

Aphrissa orbis (Poey, 1832)
Plate IV (17,18)

Recognition

FWL 26-30 mm. The cream-coloured upper surface of the male is adorned with a large, rounded, orange spot which extends from near the base of the forewing to the level of the end of the cell. The female is rich yellow-orange with a brown forewing outer margin and large cell-end spot, and brown points at the ends of the veins on the hindwing.

Subspecies
The nominate subspecies, which is established in Cuba and the Isle of Pines, has been found in the Cayman Islands. Hispaniola has an endemic subspecies.

Species' range
This is a butterfly with a range restricted to Cuba, the Isle of Pines and Hispaniola. It occurs as a vagrant in Florida and the Cayman Islands.

Cayman Islands distribution
Grand Cayman (vagrant), Cayman Brac (vagrant)

Habitat
All known specimens from Grand Cayman were found in gardens, but Smith *et al.* (1994) describe *A. orbis* as 'primarily a butterfly of upland deciduous forest, more occasionally seen at sea level and in the lowlands'.

History
On Grand Cayman, a male was found in south George Town on 25 August 1998, nectaring at *Jatropha integerrima* (Laura Stafford), and a second male was captured on an unknown date, also in south George Town (Ann Stafford). A male and female were collected on 11 March 2004 at West Bay (Askew). The female had a malformed left forewing and must have emerged on Grand Cayman, its condition probably being incompatible with a flight over the sea. The occurrence of *A. orbis* on Cayman Brac is reported by Smith *et al.* (1994) who write that it 'was collected ... recently on Cayman Brac', but no further details are given.

Biology
A larval food plant of the Orbed Sulphur is *Caesalpinia (= Poinciana) pulcherrima* (Pride of Barbados) (Fabaceae), and the caterpillar is green, shading to orange posteriorly, with yellowish tubercles and white lateral lines (Riley 1975).

Upper surface of male Orbed Sulphur caught GC (25.viii.1998) by Laura Stafford

Lower surface of the same insect

Mounted female Orbed Sulphur with malformed wing, GC (11.iii.2004)

Swallowtail Butterflies

Papilionidae

Probably the most impressive of butterflies, Swallowtails are found in every zoogeographical region. They are mostly large or very large, and many bear long hindwing tails. Characteristics of the family are the possession of six functional walking legs, each of the front pair having a spur (epiphysis) on the tibia, and the presence in the larva of an eversible, fleshy forked organ (osmeterium) on the prothorax. The osmeterium is connected to a scent gland and is everted when the larva is molested, emitting a strong defensive scent which probably drives away parasitic and predatory insects. The foodplants of most papilionid caterpillars are in families (Aristolochiaceae, Rutaceae, Piperaceae and several others) which produce toxic alkaloids, and these may be sequestered by the larvae for their own defence, and that of the adult butterflies, against predatory vertebrates.

The pupa in Papilionidae is held against the substrate by a silken girdle as well as by the posterior cremaster, and a behavioural trait of adult Swallowtails is the fluttering of the forewings when the butterflies are nectaring, a habit which seems to be unique to the family (DeVries 1987).

The three species of Caymanian Papilionidae belong to two tribes. Troidini, the tribe which includes the spectacular Birdwings of south-east Asia also contains *Battus*, a genus of mostly black Swallowtails some of which are without tails. They are considered to be generally distasteful to vertebrate predators, their larvae feeding on Aristolochiaceae and sequestering toxic aristolochic acids. In South and Central America they are often models in Batesian and Müllerian mimicry assemblages.

The second tribe is Papilionini, typical tailed Swallowtails, the Caribbean species of which are either all assigned to the genus *Papilio* Linnaeus or, as here, allocated to a few genera including *Heraclides* Hübner.

Gold Rim Swallowtail

Battus polydamas (Linnaeus, 1758)
Plate VI (1)

Recognition
FWL 42-47 mm. *Battus polydamas* is a Swallowtail Butterfly without tails, although the hindwing has a scalloped margin. It is mainly black with a greenish sheen when fresh. There is a postdiscal series of spots on both wings, yellow on the forewing, greenish yellow on the hindwing. Upper and under surfaces are similar, but the hindwing underside is paler with a small red spot at its base and a series of seven irregular, red submarginal spots. The second to fourth submarginal red spots are capped anteriorly with white, and there is a marginal series of yellow crescents in the concavities of the scalloped margin. The male has long, black sense hairs along the inner margin of the hindwing.

Subspecies
Battus polydamas cubensis (Dufrane, 1946) is the subspecies found on Grand

Cayman (Riley 1975, Smith *et al.* 1994). It occurs also in Cuba and the Isle of Pines. Subspecies of *B. polydamas* have developed freely in the West Indies, no fewer than thirteen being recognized, a situation reminiscent of the widespread formation of island races in *Dryas iulia* (page 62) resulting from very limited inter-island movement of the butterflies. *B. p. cubensis* was the most recent subspecies to be described.

Species' range
A butterfly which occurs from the southern United States (Georgia), through Central America to South America (Patagonia), and in the Bahamas and throughout the West Indies.

Cayman Islands distribution
Grand Cayman

*Male Gold Rim Swallowtail feeding on Bermuda Primrose (*Asystasia gangetica*), GC (19.i.2008), RRA*

Habitat
'It was never seen outside the George-town area. A strong flier, the butterfly was usually seen above the bush' (Carpenter & Lewis 1943). The nominate subspecies in Costa Rica is described by DeVries (1987) as a butterfly which visits flowers in gardens and hedgerows throughout the year, and is commonly seen in cities and pastures.

History
Battus polydamas is apparently an infrequent visitor to the Cayman Islands, sometimes establishing temporary breeding populations. In 1938 the Oxford University expedition found that *B. polydamas* did not appear on Grand Cayman until early June, 'and was at no time common' (Carpenter & Lewis 1943). The butterfly was not noted again in the Cayman Islands until 1995 when some were seen in late September and early October in George Town, Botabano, Lower Valley and in the Queen Elizabeth II Botanic Park. No specimen was captured in 1995 but good views were obtained. In 1999 Joanne Ross saw a Gold Rim Swallowtail at Webster's Estate in George Town and found an egg, possibly of *B. polydamas*, on cultivated *Aristolochia gigantea* in the Botanic Park. In May 2007 several *B. polydamas* were discovered by Peter Davey at Lower Valley in the Agricultural Pavilion grounds and adjacent ancient woodland, about thirty being seen on 12 May. Butterflies were still present at this locality in some numbers in February 2008. On 15 May 2007, Paul Watler noted oviposition on *Aristolochia odoratissima*. Prior to this, no native *Aristolochia* had been found in the Cayman Islands! Vines must, of course, have been present in 1938 to support the butterflies seen by the Oxford University expedition and it is unlikely that at this time ornamental *Aristolochia* species were being cultivated. Occasional sightings of *B. polydamas* in George Town in 2007 may have been of butterflies that had fed as larvae on cultivated *Aristolochia*.

Roosting Gold Rim Swallowtail, GC (11.v.2007), PD

Gold Rim Swallowtail larva on Aristolochia odoratissima, *GC (23.v.2007), JG*

Biology

Eggs are laid in clusters of up to about twenty on young stems of a range of *Aristolochia* species (Pipevines) (Aristolochiaceae); 56 species are listed as hostplants by Beccaloni *et al.* (2008). Caterpillars found on native *A. odoratissima* on Grand Cayman were successfully transferred to the ornamental *A. gigantea* (Dutchman's Pipe). To avoid overloading the foodplant, female butterflies are probably able to assess the numbers of eggs laid on a plant by previous females. The caterpillars in their early stages feed gregariously. The mature larva (fifth instar) is chocolate in colour with narrow, transverse, crimson stripes and paired tubercles, including long, mobile, fleshy pairs at front and rear ends. The average length of the life cycle is thirty days with ten being passed as a pupa (Dethier 1940). All stages can be found throughout the year.

On Grand Cayman nectaring was observed high up in a Cabbage Tree (*Guapira discolor*) and also at the herbaceous Bermuda Primrose (*Asystasia gangetica*). Feeding activity appeared to be concentrated early in the day, few butterflies being seen nectaring after 10.30 a.m. although they could be seen flying in woodland canopy at a height of 5 to 7 metres.

Andraemon Swallowtail

Heraclides andraemon Hübner, 1823
Plate VI (2,3)

Recognition

FWL nominate subspecies 32-40 mm, *H. a. tailori* 44-48 mm. The upper surface is black to dark brown with a broad longitudinal yellow band crossing both fore-

Freshly emerged Andraemon Swallowtail (Grand Cayman Swallowtail), GC (10.i.2007), JG

and hindwings. This band forks near the anterior margin of the forewing at vein 7. There are no submarginal yellow spots on the forewing (*cf. H. aristodemus*, below), but a narrow yellow bar is usually present at the apex of the cell. The scalloped edge to the hindwing has a series of yellow crescents close to the margin, the tail has an apical yellow spot, and there is an orange-red, blue and black eye-spot at the tornus. The under surface is more yellow than the upper surface, similarly patterned but on the hindwing with an additional series of submarginal blue spots and a rust-coloured mark internal to the blue mark nearest to the base of the tail. The sexes are similar.

Subspecies

Two subspecies of *H. andraemon* occur in the Cayman Islands. The form in the Sister Islands is assigned to the nominate subspecies which is also found in the Bahamas, Cuba, Isle of Pines, Turks and Caicos and Jamaica. Brown & Heineman (1972) note that insects from Cayman Brac and Little Cayman are smaller than those from Cuba and have 'a somewhat different pattern'. The Grand Cayman population represents an endemic subspecies, *H. a. tailori* (Rothschild & Jordan, 1906), a much larger insect with the yellow bar across the forewing cell absent or vestigial on the upper surface but broader than in *H. a. andraemon* below.

The longitudinal yellow band across the forewing upper surface of *H. a. tailori* is narrower than the dark brown area lying outside it, and it has a rather more extended rust-coloured mark near the base of the tail on the under surface. The two Cayman subspecies also differ in male genitalia, but the most readily appreciated distinguishing feature is their respective sizes. A third subspecies, *H. a. bonhotei* (Sharpe, 1900) is found in the Bahamas, Turks and Caicos Islands and, as a recent colonist, in the Florida Keys.

Species' range

From the Florida Keys, where it may be only a temporary colonist, to the Bahamas, Turks and Caicos Islands, Cuba and the Isle of Pines, Cayman Islands and, since 1945, Jamaica.

Cayman Islands distribution

Grand Cayman (*H. andraemon tailori*, the Cayman Swallowtail), Little Cayman and Cayman Brac (*H. andraemon andraemon*)

Habitat

Heraclides andraemon is a butterfly that is most frequently seen in the vicinity of light woodland, often in parks and gardens. *H. a. tailori* can occur wherever citrus trees are grown on Grand Cayman, and we have frequently seen it at the eastern end of the island, a part from which it was absent in 1938 (Carpenter & Lewis 1943).

History

Found quite commonly on all three Cayman Islands in 1938 (Carpenter & Lewis 1943), the status of *H. andraemon* does not appear to have changed since. The description of the Grand Cayman

subspecies *H. a. tailori* by Rothschild & Jordan (1906) seems to be the first mention in the literature of a butterfly in the Cayman Islands. The description is based on a male and female collected in 'Great Cayman Island' in April 1896 by 'Mr Taylor'. The collector was Charles B. Taylor and the subspecies is named for him; the spelling change is to latinize his name. Taylor lived in Jamaica and was zoological curator in the Jamaica Institute of Kingston towards the end of the nineteenth century. He collected on Grand Cayman from 14 March until 21 April 1896. The Grand Cayman endemic subspecies of the Cuban Bullfinch, *Melopyrrha nigra taylori* (Hartert, 1896), is also named after him.

Freshly emerged Andraemon Swallowtail, GC (10.i.2007), JG

Biology

The larva of *H. andraemon* feeds on plants of the family Rutaceae. *Citrus aurantifolia* (Lime Tree) is the main food-plant on Grand Cayman. *Citrus*, however, is an introduced plant, and *H. andraemon* must have been on Grand Cayman before the introduction of *Citrus* for it to have had time to evolve into a distinct subspecies. Native species of Rutaceae probably provided the original larval food source, and caterpillars have been found on Grand Cayman eating *Amyris elemifera* (Candlewood), *Zanthoxylum coriaceum* (Shake Hand Tree), *Z. fagara* (Wild Lime) and *Z. flavum* (Satinwood, Yellow Sanders). Satinwood is rare on Grand Cayman and a mature tree was found only after Carla Reid (personal communication) was led to it by a Swallowtail. Eggs are green and spherical, laid singly on leaf-tips and reported to be devoured in quantities by ants (Brown & Heineman 1972). The fully grown larva is, like other Swallowtail caterpillars, very swollen at the third thoracic segment. It bears a resemblance to a large bird dropping being grey with a brown head, red-brown mottlings, white lateral lines, a cream-coloured saddle and lilac underparts.

The Andraemon Swallowtail flies rapidly but usually at a height of only one or two metres. It nectars at a range of flowering shrubs and herbaceous plants; *Asclepias curassavica*, *Bauhinia divaricata*, *Caesalpinia pulcherrima*, *Jatropha integerrima* and *Lantana camara* are noted as being visited on Grand Cayman. When feeding, the butterfly is never still, constantly fluttering its forewings and quickly moving to the next blossom.

Dusky Swallowtail

Heraclides aristodemus (Esper, 1794)
Plate VI (4)

Recognition
FWL 41-43 mm. (Little Cayman specimens are small; elsewhere in the species'

distribution FWL ranges from 45 to 53 mm.). *Heraclides aristodemus* can be distinguished from *H. andraemon* (above), with which it flies on Little Cayman, by a series of yellow submarginal lunules on the upper surfaces of both the fore- and hindwings. The yellow crescentic marginal spots on the hindwing are smaller in *H. aristodemus* than in *H. andraemon*, and the black hindwing tail is without a yellow spot. The yellow bar crossing both wings runs more obliquely in *H. aristodemus* than in *H. andraemon*, more basal in position on the hindwing and running directly towards the forewing apex, forking more widely from a yellow spot between veins 6 and 7. This latter spot is much broader in *H. aristodemus* than in *H. andraemon*, being the broadest in the series, and it encloses a small brown mark. Most of the blue spots on the hindwing under surface have an internally adjacent rust-coloured mark.

Subspecies
Heraclides aristodemus temenes (Godart, 1819), known from Cuba, Isle of Pines and Little Cayman, has a broader yellow bar across the upper surfaces of both wings than have other subspecies. The name Dusky Swallowtail seems rather inappropriate.

Species' range
The butterfly is known from southern Florida (as *H. a. ponceanus* (Schaus, 1911), Schaus' Swallowtail, an officially listed endangered subspecies, now restricted to the Keys), the Bahamas, Turks and Caicos Islands, Cuba, Little Cayman, Hispaniola, Mona Island and Puerto Rico.

Cayman Islands distribution
Little Cayman

Habitat

C. B. Lewis in Carpenter & Lewis (1943) writes that *H. aristodemus* 'was not observed on the low hills or in the eastern part of the island [Little Cayman] although, it must be admitted, these sections were poorly explored owing to the virtually impenetrable bush'. It frequents paths and tracks, and unlike many other butterflies on Little Cayman, it is not confined to dense vegetation. Elsewhere in its range, *H. aristodemus* is an insect of low lying, dry scrub and woodland, sometimes coming into gardens.

History

'*Papilio aristodemus temenes* was one of the less common species' when discovered on Little Cayman in 1938 (Carpenter & Lewis 1943). In 1975 only two were seen during six weeks of exploration of the island, a female at Pirates Point on 17 July and a male north of Blossom Village on 2 August (Askew 1980). In 2008 a single Dusky Swallowtail was seen on 25 January on the north coast of Little Cayman. The species has never been found on Cayman Brac.

Biology

The larval food-plants in Florida and Cuba are *Amyris elemifera* (Candlewood, Torchwood), *A. balsamifera* and, less commonly, *Zanthoxylum fagara* (Wild Lime) (Rutaceae). The small, spherical, cream-coloured eggs are laid singly on leaf-tips. When fully grown the larva is brown, with irregular cream-coloured blotches and blue dots on its back and sides. The pupa is described as twig-like (Smith *et al.* 1994).

The flight of the Dusky Swallowtail is described as less direct than that of the Andraemon Swallowtail, but just as rapid, and activity peaks late in the morning with sometimes a second burst just before dusk (Smith *et al.* 1994). Several nectar plants have been recorded outside the Cayman Islands, including *Bauhinia divaricata, Bidens alba, Cordia, Lantana* and *Tournefortia*.

Skippers

Hesperiidae

Hesperiidae are so different from other butterflies that they are put in a super-family of their own, Hesperioidea, all other butterflies being in the superfamily Papilionoidea. The adult Skipper has no ocelli (simple eyes) between the compound eyes, and its antennae are widely separated, with the apical clubs gradually, not sharply, expanded and hooked at their tips. The tibial spur on the front leg (the epiphysis) is much modified, adpressed and minutely pubescent, with a covering of long scale hairs, and the hind legs usually bear two pairs of tibial spurs. The colour of the wings is usually dull, mostly brown and black, often with whitish, translucent markings, and the forewings are relatively short and narrow. The body is stout. Flight in Skippers, as befits their name, is rapid and darting.

Two subfamilies of Hesperiidae are represented in the Cayman Islands: Pyrginae which includes *Phocides* and *Urbanus*, and Hesperiinae which includes *Asbolis*, *Calpodes*, *Cymaenes*, *Hylephila* and *Panoquina*. When at rest the wings of Pyrginae are held horizontally, and there is usually a costal fold along the leading edge of the male forewing. In Hesperiinae the resting butterfly usually adopts a characteristic posture with the hindwings almost horizontal but the forewings only partly open, and the forewing of the male often has an oblique dark band of andro-conial scales.

Larvae of Pyrginae feed on dicotyledonous plants, whereas those of Hesperiinae eat monocotyledons, especially grasses. The head of the caterpillar in Pyrginae is described as monkey-faced because of its prominent eye-spots.

Skipper caterpillars are usually yellowish to green with pale lateral lines. The head is large and set off from the rest of the body by a constriction or neck. Fully grown larvae, and pupae, are often covered with a white, waxy and powdery exudate. The larvae construct shelters of leaves bound with silk, and they usually pupate in these shelters.

Mangrove Skipper

Phocides pigmalion (Cramer, 1779)
Plate VI (5,6)

Recognition
FWL 22-27 mm. This is an unmistakable butterfly, the upper surface dark brown with iridescent blue stripes and a submarginal series of blue spots on the

Mangrove Skipper feeding at Sea Lavender (Argusia gnaphalodes), LC (23.i.2008), RRA

Mangrove Skipper feeding at Sea Lavender, LC (23.i.2008), RRA

hindwing, and blue marks at the base of the forewing. The under surface is similar but has rather broader blue stripes on the hindwing. There are also blue longitudinal stripes on the thorax, paired blue spots on the abdomen, and the undersides of the palps, either side of the proboscis, are white. The sexes are similarly patterned but the male is smaller and has a fold in the forewing, just behind the costal margin, which runs from near the base of the wing to about two-thirds of the way to the apex. This costal fold contains androconial scent scales.

Subspecies

Phocides pigmalion batabano (Lucas, 1857) is known from Andros in the Bahamas, Cuba, the Isle of Pines and Little Cayman. In lacking pale, hyaline forewing markings, it differs strikingly from *P. p. bicolora* (Boddaert, 1783) from Hispaniola, and other subspecies from South America, but resembles the forms flying in Florida and the Bahamas.

Species' range

The Mangrove Skipper is found from southern Florida, through Central America to South America (Argentina), and in the Bahamas, Cuba and the Isle of Pines, Little Cayman, Hispaniola and, at one time, Puerto Rico (now probably extinct).

Cayman Islands distribution
Little Cayman

Habitat
C. B. Lewis in Carpenter & Lewis (1943) writes: 'The trip across Salt Rocks Hill [Little Cayman] was a hard battle through the bush. We were probably the first to cross that section since the days of the pirates. Directly at the top of the hill we caught three specimens [of *P. pigmalion batabano*] ... Capturing this form is a real task as they fly above the trees and alight in the foliage at considerable heights'. In 1975 both sexes were found flying high up about a row of *Rhizophora mangle* (Red Mangrove) growing at the sea's edge, and in 2008 many were observed feeding at the blossoms of sweet-scented Sea Lavender on beach ridges.

History
This butterfly was found on Little Cayman on 31 May 1938 at Salt Rocks Hill (above) (Carpenter & Lewis 1943) and in 1975 at East Rocky Point (Askew 1980). It was first seen at the last named locality on 11 July, but a voucher specimen could not be captured until 25 July. Frank Roulstone found the Mangrove Skipper quite plentifully on Little Cayman in 2005 and in 2008 several were seen by the shore between East Rocky Point and Charles Bight on Little Cayman's south coast. The Mangrove Skipper remains unrecorded from Cayman Brac and Grand Cayman.

Biology
The egg is almost spherical with vertical ridges, and laid on *Rhizophora mangle* (Red Mangrove) (Rhizophoraceae). *Rhizophora* grows in shallow sea water and is a well known larval food-plant, but the occurrence of the butterfly at high altitude away from the coast in other parts of the species' range, for example at 1400m altitude in the Sierra Maestra of Cuba (Hernández 2004), suggests that there are alternative food-plants. Recently, in fact, the caterpillar has been found in Hispaniola feeding on *Melicoccus bijugatus* (Ginep) (Sapindaceae), a naturalized Cayman plant. The fully grown caterpillar is stout and reddish brown with transverse yellow bands, but the markings are obliterated by a white, powdery exudate.

Although often seen flying high about trees, the butterfly also takes nectar from flowers of *Bauhinia divaricata*, *Bidens alba* and *Caesalpinia* (Smith *et al.* 1994). The butterflies seen in 1975 appeared to be feeding at flowers of *Rhizophora mangle*. Other nectar sources on Little Cayman are *Stachytarpheta jamaicensis* and especially Sea Lavender (*Argusia gnaphalodes*).

Long-tailed Skipper

Urbanus proteus (Linnaeus, 1758)
Plate VI (7)

Recognition
FWL 23-26 mm. This, like *Phocides pigmalion*, is quite a large skipper and easily recognizable by the hindwing tails and blue-green hairs on the body and bases of the wings. There are white, semitransparent spots on the apical half of the forewing. These spots are present on upper and under wing surfaces. The under surface is brown with dark brown spots and

bars on the hindwing. The sexes are quite similar although the semi-transparent forewing spots are smaller in the male than in the female, and the male has a prominent costal fold.

Subspecies
Urbanus proteus domingo (Scudder, 1872) flies in the Bahamas and throughout the West Indies, but it is only weakly differentiated from the nominate form, chiefly by its reduced white markings.

Species' range
This is a very widely distributed skipper, occurring from the southern United States through Central America and south to Argentina. It is found throughout the West Indies.

Cayman Islands distribution
Grand Cayman, Cayman Brac

Habitat
The Long-tailed Skipper is a butterfly of open, sunny places, flying in gardens, meadows and uncultivated areas with moderately short vegetation.

Female Long-tailed Skipper, GC (28.xii.2005), JG

*Long-tailed Skipper taking nectar from flower of Vervine (*Stachytarpheta jamaicensis*), GC (15.i.2008), RRA*

History
This butterfly has been assessed as locally frequent to widespread and common on Grand Cayman since it was first recorded there in 1938. It was also found on Cayman Brac in 1938 (Carpenter & Lewis 1943), and again in 1990 (Miller & Steinhauser 1992). With regard to Little Cayman, Lewis in Carpenter & Lewis (1943) writes 'While no specimens were collected or seen on Little Cayman [in 1938], I expect that it occurs'; it was not found on Little Cayman during the six-week expedition in 1975.

Biology
The larva of *Urbanus proteus* is classified as a minor pest, feeding on a wide range of plants, but especially wild legumes and cultivated beans and peas. The list of larval food-plants includes *Clitoria, Desmodium, Phaseolus* (= *Macroptilium*), *Vigna* and *Wisteria* (Fabaceae), *Melochia* (Ster-

culiaceae) and *Brassica* (Cruciferae). On Grand Cayman, caterpillars have been found on *Desmodium tortuosum* (Dixie Tick Trefoil) and *Melochia tomentosa* (Velvet Leaf). They live in leaf-rolls and may be parasitized by a braconid wasp (*Cotesia* (as *Apanteles*)) (Brown & Heineman 1972). When fully grown, the caterpillars are dull green speckled with black and have a fine, black mid-dorsal line and yellow lateral lines; there are orange eye-spots on the brown head.

Urbanus proteus flies rapidly and erratically but perches frequently in prominent places. Males appear to be territorial and return repeatedly to a favoured perch after making sorties at passing insects. A large proportion of butterflies lose one or both tails, probably as a result of attacks by lizards. A variety of nectar plants are used, including *Bidens alba*, *Bougainvillea*, *Lantana camara*, *Stachytarpheta jamaicensis* and *Tournefortia*. Diurnal flight activity is prolonged, commencing soon after sunrise and continuing until dusk or later. The butterflies are sometimes attracted to electric lights.

Dorantes Skipper, Brown-tailed Skipper

Urbanus dorantes (Stoll, 1790)
Plate VI (8)

Recognition
FWL 21-23 mm. This butterfly resembles *Urbanus proteus* (above), but it lacks blue-green iridescent pilosity on the upper surface and has smaller white spots on the forewing. The hindwing tails of *U. dorantes* are slightly shorter than those of the Long-tailed Skipper, and they project at a smaller angle, not much over 90°, to the outer edge of the wing.

Subspecies
Urbanus dorantes santiago (Lucas, 1857), a subspecies from the Bahamas, Cuba and the Isle of Pines, has been found infrequently on Grand Cayman, and the occurrence on the island of another subspecies, *U. d. cramptoni* Comstock, 1944 from Hispaniola and further east has also been reported (see below under History). *U. d. santiago* has a darker hindwing under surface than *U. d. cramptoni*, the dark bands being broader and interspersed by brown, not purplish, ground colour, and the outer margin is grey-brown with a paler grey band running between it and the distal dark brown band.

Species' range
The distributions of both species of *Urbanus* considered here are quite similar, the Dorantes Skipper being known from the southern United States through Central America and in South America south to Argentina, and in the Bahamas, Cuba, Grand Cayman, Hispaniola, Mona Island, Puerto Rico and into the Virgin Islands.

Cayman Islands distribution
Grand Cayman

Habitat
Urbanus dorantes is found in similar places to *U. proteus*, the two species often flying together (Smith *et al.* 1994).

History
The first published record of *U. dorantes* in the Cayman Islands is by Schwartz *et al.* (1987) who captured ten specimens, which they attribute to *U. d. cramptoni*, at the end of November 1985 at Boat-

swain Bay, Cayman Kai and Old Man Bay on Grand Cayman. These authors suggest that this broad distribution on the island indicates that the species is 'not an extremely recent adventive, although it must have arrived since Askew (1980) collected there in 1975'. The record of Schwartz *et al.* (1987) is, in fact, pre-dated by the discovery of a specimen on 28 December 1975 by Dr E. J. Gerberg (pers. comm.). It was found later, on 17 August 1985 when a single fresh male of *U. d. santiago* was taken on a dyke road north of South Sound (Askew), and in February 1987 when Dr J. Clarke collected an example at an unknown locality in Grand Cayman (Askew 1988). Finally, a female was captured at Great Beach on 2 November 1990 (Miller & Steinhauser 1992).

It should be noted that Schwartz *et al.* (1987), who have examined the largest number of Caymanian specimens, attribute their insects to *U. dorantes cramptoni*, whereas other authors identify the subspecies they found as *U. dorantes santiago*. It seems unlikely that two different subspecies would occur at virtually the same time on Grand Cayman, and it would be desirable to have the subspecific identity confirmed of all Grand Cayman material of *U. dorantes*.

Biology

Larval food-plants are a large range of legumes including *Clitoria, Desmodium* and *Phaseolus* (= *Macroptilium*).

The butterflies have an erratic, darting flight like that of the Long-tailed Skipper (above), and the two species have a similar range of nectar plants (Smith *et al.* 1994).

Three-spotted Skipper, Dingy Dotted Skipper

Cymaenes tripunctus (Herrich-Schäffer, 1865)
Plate VI (9)

Recognition

FWL 14-15 mm. This is a very plain little Skipper, mostly dark brown with a few inconspicuous, whitish, semitransparent spots on the forewing. Two of these spots are central (discal), and three very small ones (the most anterior often indistinct) are in an oblique row near the apical margin. The latter character gave rise to the vernacular name of Three-spotted Skipper, but it should be noted that the name is shared with an unrelated North American Skipper, *Oligoria maculata* (Edwards, 1865) (Gerberg & Arnett 1989). The apex of the forewing in *Cymaenes tripunctus* is not sharply pointed and the tornus of the hindwing is not extended. This last character is a good field guide for distinguishing *Cymaenes* from *Panoquina panoquinoides* (below). *C. tripunctus* has a more gradually expanded antennal

Three-spotted Skipper in typical hesperiine resting posture, GC (6.ii.2006), RRA

club than *P. panoquinoides*, and the bent-over apical process (apiculus) is distinctly longer than the breadth of the club (Figure 3), and the angle between apiculus and the body of the club is greater in *C. tripunctus* than in *P. panoquinoides*. Other characters which distinguish *C. tripunctus* from *P. panoquinoides* are its darker brown ground colour, its white rather than yellowish wing spots and head scales (especially below the eye) and relatively longer antennae which are more than half the length of the forewing. The sexes of *Cymaenes* are very similar.

Three-spotted Skipper GC (6.ii.2006), RRA

Subspecies
The nominate subspecies is West Indian. Another subspecies is found on the mainland (except Florida).

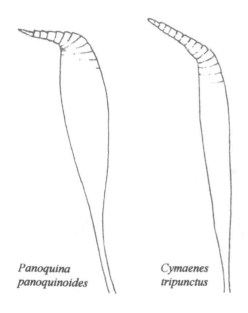

Panoquina
panoquinoides

Cymaenes
tripunctus

Figure 3. Clubs of left antennae (dorsal view) of Panoquina panoquinoides *and* Cymaenes tripunctus.

Species' range
Found from the southern United States (Florida), the Bahamas, Greater Antilles and associated islands eastwards to the Virgin Islands and, as *C. t. theogenis* (Capronnier, 1874), from Mexico through South America to Argentina.

Cayman Islands distribution
Grand Cayman, Cayman Brac

Habitat
The Three-spotted Skipper, like others of its family, is a butterfly of open places with rank, herbaceous vegetation, occurring in uncultivated areas and also in parks and gardens.

History
Found commonly on Grand Cayman in 1938 by the Oxford University expedition, *Cymaenes tripunctus* has since been moderately frequent in most years on the island. It was discovered on Cayman Brac in 1975 (Askew 1980) and found again in 1990 (Miller & Steinhauser 1992), locally common in both years. There are no records for Little Cayman.

Biology

Eggs are laid singly on grass blades, and the larvae construct a series of leaf-fold shelters. The pupa is also in a partly folded leaf (Brown & Heineman 1972). Larval food-plants include the grasses *Bambusa vulgaris* (Bamboo), *Panicum maximum* (Guinea Grass) and *Saccharum officinale* (Sugar Cane) (Poaceae), but its diet in the Cayman Islands is unknown. The caterpillar is bluish green with a dark mid-dorsal stripe and pale green lateral stripes.

This is a low-flying but inconspicuous butterfly. It has been observed nectaring at flowers of *Lantana camara* and *Stachytarpheta jamaicensis* on Grand Cayman, but spends long periods perched on foliage with wings held in the characteristic posture of a hesperiine (above). After dark, it is occasionally attracted to electric lights.

Fiery Skipper

Hylephila phyleus (Drury, 1773)
Plate VI (10,11)

Recognition

FWL 15-17 mm. Unlike the other Skippers in the Cayman Islands with the exception of the Striped Skipper, the sexes of the Fiery Skipper are quite differently patterned. The male is bright fulvous on the upper surface with an oblique, black sex brand across the forewing cell, a dark brown, sinuate bar at the end of the forewing cell and dark brown, mostly triangular, marginal marks on the outer wing margins. The under surface is similar but the dark marks are reduced. The female is much darker than the male, the dark brown markings on the upper surface being expanded and restricting the bright fulvous ground colour to little more than two rows of spots.

Subspecies

The nominate subspecies occurs in the West Indies.

Species' range

Hylephila phyleus ranges from Connecticut and Iowa in the United States, south through Central America to South America reaching Argentina, and throughout the West Indies.

Cayman Islands distribution

Grand Cayman, Cayman Brac

Habitat

The Fiery Skipper flies in open, grassy places with short and often sparse vegetation cover. It is often found on beach ridges in the Cayman Islands.

History

Hylephila phyleus was found in 1938 on Grand Cayman and Cayman Brac (Carpenter & Lewis 1943), but it was not seen on any of the Cayman Islands in 1975

Male Fiery Skipper, GC (6.ii.2008), RRA

Female Fiery Skipper on Cosmos sp.*, GC (24.vii.2005), AS*

(Askew 1980). In more recent times, *H. phyleus* has been localized and never plentiful on Grand Cayman. Schwartz *et al.* (1987) repeat the 1938 distribution but do not mention finding the butterfly themselves when they visited Grand Cayman and Cayman Brac in 1985. Miller & Steinhauser (1992) found the species in 1990 on Grand Cayman but not on Cayman Brac.

Biology

The smooth-surfaced yellow-green eggs are laid singly on grass blades. No larval food-plant has been reported for the Cayman Islands, but elsewhere the caterpillars are known to feed on grasses in the genera *Cynodon, Digitaria, Panicum, Paspalum, Saccharum* and *Stenotaphrum*. The young caterpillar, which is greenish, binds a grass blade with silk so that it is partly folded. It feeds both at night and during day-time, leaving its shelter to eat but retreating backwards into it if disturbed. Older larvae construct larger shelters incorporating a number of grass blades, but when about half grown the larva loses its green colour and becomes brown with darker longitudinal stripes. It descends to the ground and excavates a burrow into which it pulls, mainly at night, pieces of grass on which it feeds. Pupation takes place in the burrow (Brown & Heineman 1972).

Male butterflies perch in full sunlight on low vegetation or even bare earth and appear to be territorial. When flying, both sexes keep close to the ground. They take nectar from a variety of herbaceous and shrubby plants; known nectar sources on Grand Cayman are *Bidens alba, Lantana camara* and *Stachytarpheta jamaicensis*, and on Cayman Brac *Argusia gnaphalodes* and *Suriana maritima*.

Pair of Fiery Skippers in copulation, male on the right, GC (29.i.2006), RRA

Striped Skipper

Atalopedes mesogramma (Latreille, 1824)
Plate VI (12,13)

Recognition

FWL 18-20 mm. The male is brown with the wings fulvous basally, the forewing with an elongated dark sex brand and subapical fulvous spots. It somewhat resembles a large, dark *Hylephila phyleus*. The female is medium brown without fulvous basal areas to the wings and the subapical

Female Striped Skipper, LC (8.xii.2007), FR

Male Striped Skipper, CB (30.i.2008), RRA

forewing spots are white. In both sexes the Striped Skipper is characterized by a white band of postmedial spots across the under surface of the hindwing.

Subspecies
The three Cayman specimens found are attributed to the nominate subspecies. This occurs in the Bahamas and Cuba, and is replaced in Hispaniola and Puerto Rico by a smaller and brighter form, *A. m. apa* Comstock, 1944.

Species' range
A north Caribbean butterfly flying in the Bahamas, Cuba (widespread and common), Hispaniola and Puerto Rico, and on the Yucatan Peninsula.

Cayman Islands distribution
Little Cayman, Cayman Brac (vagrant?)

Habitat
The Striped Skipper occurs in both moist, open grassy places and very dry, open scrub (Smith *et al.* 1994). In Cuba it is common in gardens, roadsides, lawns and the like (Hernández 2004). All three Cayman Islands' records are from, or close to, the coast.

History
The Striped Skipper is the most recent addition to the list of Cayman Islands' butterflies. A female *A. mesogramma* was discovered and photographed on Little Cayman at the airstrip end of the nature trail by Frank Roulstone on 8 December 2007. Another female, only the wings and thoracic exoskeleton remaining, was found on the floor of a house at West End, Cayman Brac in January 2008, and a male was located on the beach ridge at Creek, Cayman Brac on 30 January 2008 (Askew).

Biology
The caterpillar of the Striped Skipper feeds on *Cynodon dactylon*, *Eleusine indica*, *Paspalum fimbriatum*, *Sorghum* and other grasses, and when fully grown it is yellow-green with brown head and legs, the body covered with bristly hairs. The adult insect has been seen nectaring at *Borrichia arborescens* and *Suriana maritima* on Cayman Brac, and elsewhere it is reported to feed at *Ageratum conyzoides*, *Bidens alba*, *Bougainvillea* and *Lantana* (Smith *et al.* 1994).

Monk

Asbolis capucinus (Lucas, 1857)
Plate VI (14)

Recognition
FWL 20-26 mm. This is a large brown Skipper, darker above than below, without any whitish markings except, in the male, a very thin and oblique, pale, discal streak on the forewing. The upper surface of the hindwing is a little darker than that of the forewing, and the forewing apex is produced. The sexes are similar.

Female Monk on Pygmy Date Palm (Phoenix roebelenii)*, GC (11.vi.2007), AS*

Subspecies
None has been described.

Species' range
The Monk has a limited distribution. It was known only from Cuba and the Isle of Pines until about 1947, when it colonized Florida, and in 1981 it appeared in the Bahamas (New Providence) (Smith *et al.* 1994). We first found it on Grand Cayman in 2002.

Cayman Islands distribution
Grand Cayman

Habitat
Found especially in the neighbourhood of native and cultivated palms, the larval food-plants, and frequently a garden butterfly.

History
The first specimen of this mainly Cuban insect was taken in a south George Town garden on 10 March 2002 (Askew) and further specimens have since been seen fairly regularly in the same garden. In March 2004 it was found in Bodden Town and East End, and in 2006 at Bodden Town in January and Heritage Beach in March. The status of the Monk on Grand Cayman at the time of writing can be described as rare but widespread. It will be interesting to see if it becomes established here as it did in Florida.

Biology
The larvae of *A. capucinus* feed on a broad range of palms, including *Cocos nucifera* (Coconut Palm) and *Phoenix* species (Date Palms). An adult visited *P. roebelenii* (Pygmy Date Palm), a very likely larval food-plant, in a garden in George Town on several days in June 2007. Other palms that could serve as food-plants in the Cayman Islands are listed in the table of larval food-plants and nectar flowers (page 148).

Adult butterflies are very wary and fly fast. They remain active until well after sunset like many other Skippers, and will fly to electric lights. Nectar is obtained from flowering herbaceous plants and shrubs, a very long proboscis enabling the butterflies to reach nectar in tubular blossoms. *Hibiscus, Lantana camara, Turnera ulmifolia* and *Stachytarpheta jamaicensis* are visited on Grand Cayman.

Canna Skipper, Brazilian Skipper

Calpodes ethlius (Stoll, 1782)

Plate VI (17)

Recognition

FWL 24-26 mm. This is a large brown Skipper with semitransparent white markings on both front and hindwings, and no tails. The ground colour of the upper surface is dark brown, brighter and more golden brown towards the wing bases. The under surface is mostly yellow-brown with white markings as on the upper surface. The sexes are similar. In wing-shape *Calpodes* resembles *Panoquina* (below) in having the apex of the forewing produced, but *C. ethlius* may be distinguished from the much more numerous *P. lucas* by its greater size and generally larger white markings with a conspicuous line of three pale spots on the hindwing upper surface, and golden brown bases to the wings.

Subspecies

Uniform in appearance throughout its wide range, no subspecies have been described.

Species' range

Calpodes ethlius occurs in the United States, breeding from South Carolina and Texas southwards and occasionally straying as far north as New York, through Central America and the West Indies, south to Argentina.

Cayman Islands distribution

Grand Cayman

Habitat

This butterfly flies in parks and gardens, usually in the vicinity of the larval food-plants, various species of *Canna*, where these are grown as ornamentals (Canna Lilies). It is a highly dispersive insect.

Reared female Canna Skipper, GC (10.ii.2008), MLA

Larva of Canna Skipper in opened leaf-roll, GC (7.ii.2008), MLA

Pupa of Canna Skipper in partly opened Canna leaf-roll, GC (7.ii.2008), MLA

History

The Canna Skipper is mentioned as a Caymanian insect by Riley (1975), but it was not found in 1938 by the Oxford University expedition (Carpenter & Lewis 1943). A specimen taken in George Town, Grand Cayman about 1971 was seen in the collection of J. Francois Lesieur (Askew 1980), and more recently a pair in the National Trust reference collection was reared by Joanne Ross in 2003 from larvae found on *Canna* purchased at a George Town nursery garden. A specimen seen in a demonstration collection in the Agricultural College was collected as a larva feeding on *Canna* by Joan Steer at Lower Valley in 2005, and in 2008 larvae and pupae were found in leaf-folds of Canna Lilies in the Dart Arboretum, George Town.

Biology

The caterpillar of *C. ethlius* feeds on various species of *Canna* (e.g. *C. flaccida*, *C. indica*), and in places where these lilies are grown ornamentally, such as Miami and Jamaica, it may become a horticultural pest. The greenish caterpillar is elongated and semitransparent. It constructs a shelter from a rolled or folded *Canna* leaf bound with silk threads. The pupa, which also is in a partly folded leaf, is about 4 cm. long, thin, green and with a white, waxy exudate. It has a brown spike at the anterior end and a proboscis case which extends back well beyond the tip of the abdomen. The pupa is held in position by the cremaster locked into a silken pad and by a silken girdle about the thorax. The empty pupal case is almost colourless. Caterpillars are heavily parasitized by parasitic wasps (Hymenoptera, Ichneumonoidea) and flies (Diptera, Tachinidae) (Brown & Heineman 1972).

Sugar Cane Skipper

Panoquina lucas (Fabricius, 1793)
Plate VI (15)

Recognition

FWL 17-18 mm. The wing shape of the Sugar Cane Skipper is similar to that of the Canna Skipper (above) with an extended forewing apex and produced tornal area on the hindwing, but *P. lucas* is a smaller insect than *C. ethlius* with reduced white markings, those on the hindwing no more than a very faint, oblique submarginal row of 7 or 8 small, lilac-white spots on the under surface. There is a white chevron near the centre of the forewing. Other species of hesperiine Skippers in the West Indies resemble *P. lucas* and can be distinguished, even in the hand, only with difficulty. These include *P. ocola* Edwards and *Nyctelius nyctelius* (Latreille), both reported from the Cayman Islands but whose presence here needs confirmation (see under Reputed Caymanian butterflies; both lack the oblique row of small pale spots on the hindwing under surface, a faint but diagnostic feature of *P. lucas*).

Subspecies
Carpenter & Lewis (1943) record this species from Grand Cayman under the name *Panoquina sylvicola* (Herrich-Schäffer, 1865), the material collected by the 1938 Oxford University expedition stated to include both *P. l. woodruffi* (Watson, 1937) (23 specimens) and *P. l. lucas* (as *P. sylvicola sylvicola* (Herrich-Schäffer, 1865)) (2 specimens). The purported differences between these subspecies are extremely slight and we prefer to assign all Caymanian insects to the nominate subspecies which flies on the mainland and in Cuba. *P. l. woodruffi* occurs on Jamaica, Hispaniola and other West Indian islands.

Sugar Cane Skipper nectaring at Sea Lavender (Argusia gnaphalodes), GC (26.i.2008), TP

Species' range
This is a wide-ranging butterfly, occurring from Texas through Central America, in South America south to Argentina, and throughout the West Indies (but absent from the Bahamas).

Cayman Islands distribution
Grand Cayman

Habitat
Panoquina lucas flies almost anywhere on Grand Cayman but avoids dense woodland and mangrove swamps. It is a frequent visitor to parks and gardens, and common also on the beach ridges and in open scrubland. The junior synonym *sylvicola* means 'a woodland dweller'.

History
A generally plentiful insect on Grand Cayman since first found there in 1938 Carpenter & Lewis 1943), but not recorded from Little Cayman or Cayman Brac.

Sugar Cane Skipper, GC (31.i.2006), RRA
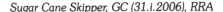

Biology

The caterpillars of *P. lucas* feed on various grasses, including Bamboo (*Bambusa vulgaris*), Rice (*Oryza* species), Sorghum (*Sorghum halepense*) and Sugar Cane (*Saccharum officinale*). They may attain pest status on the latter crop in Puerto Rico (Brown & Heineman 1972). Larval food-plants on Grand Cayman have not been ascertained. Eggs are laid singly on grass blades and the green larvae shelter in a leaf-fold during the day, feeding at night. The pupa is also formed in a leaf-fold, attached to a silken pad at the posterior end and by a silken girdle about the thorax.

The butterflies have been observed feeding at *Stachytarpheta jamaicensis* on Grand Cayman, but undoubtedly a wide range of flowers provide them with nectar.

A number of parasitic wasps attack the immature stages of *P. lucas* in Puerto Rico (Jones & Wolcott 1922). Larvae of an *Apanteles* or *Cotesia* species (Braconidae) feed gregariously inside the developing caterpillar, chewing their way out of the body of their host only when they are fully grown, and spinning their white cocoons *en masse* about the still living but shrunken host body. *Ardalus* (Eulophidae) is also gregarious (up to sixteen individuals per host), but its larvae feed outside rather than inside the host caterpillar and they form exposed black pupae about the host remains. The egg parasitoids *Ooencyrtus* (Encyrtidae), which are solitary, and *Trichogramma* (Trichogrammatidae) which are gregarious, seven to eleven individuals developing in a single butterfly egg, are also recorded.

Obscure Skipper

Panoquina panoquinoides (Skinner, 1891)
Plate VI (16)

Recognition

FWL 13-14 mm. The Obscure Skipper is like a smaller version of the Sugar Cane Skipper (above), with similarly shaped, brown wings. *P. panoquinoides,* however, is a lighter shade of brown than *P. lucas* and the pale yellowish spots on the forewing are very small. The central chevron is short and almost triangular. In flight, *P. panoquinoides* could be mistaken for *Cymaenes tripunctus* (above) if wing shape is not clearly seen. The apex of the forewing and the tornal area of the hindwing of *Panoquina* are produced, but the wings of *Cymaenes* are shorter and more rounded. These differences are most easily appreciated in the perching insects. The antennal club (Figure 3) is more sharply expanded in *P. panoquinoides* than in *Cymaenes tripunctus* and the apiculus is strongly bent over, forming almost a right angle to the body of the club. The apiculus is shorter than the breadth of the club (longer in *C. tripunctus*). The wings of *Panoquina* are a warmer shade of brown than those of *Cymaenes,* and there are only two (not three) small, apical forewing spots. The pale scales on the head are buff, not white.

Subspecies

The nominate subspecies, *P. p. panoquinoides* (Skinner), flies in the Florida Keys and in the West Indies south-east to St Lucia. It is replaced in the Grenadines and Grenada by *P. p. eugeon* (Godman & Salvin, 1896), which is considered by Riley (1975) to be perhaps a distinct species.

Obscure Skipper, CB (31.i.2008), RRA

Species' range

Panoquina panoquinoides is distributed from the southern United States, through Central America and in South America as far south as Brazil and Peru. It is found in the Bahamas, and is widespread in the Caribbean from Cuba to Grenada although occurring somewhat erratically and unrecorded from a number of the Lesser Antilles (Smith *et al.* 1994).

Cayman Islands distribution

Grand Cayman, Little Cayman, Cayman Brac

Habitat

The Obscure Skipper is mainly a coastal species, inhabiting the bushy vegetation of the beach ridges and margins to the brackish lagoons. Lewis in Carpenter & Lewis (1943) describes it as being found 'only within a hundred yards of the sea in the scrubby vegetation growing at the top of and behind the beaches of the north and east coasts of Grand Cayman. On the Sister Islands it was taken in pasture lands much further from the sea'. Hernández (2004) remarks on its association with very saline habitats in Cuba.

History

The apparently erratic and irregular occurrence of *P. panoquinoides* over much of its West Indian range is recounted by Smith *et al.* (1974), and is exemplified by its history in the Cayman Islands. It was found on all three islands in 1938 (Carpenter & Lewis 1943) and 1975 (Askew 1980), when it was local but not rare. Inexplicably, it seems to have disappeared from Grand Cayman but it was still present, although only in small numbers, on the Sister Islands in 2008.

Biology

The larva of *P. panoquinoides* feeds on grasses. It has been reared on *Cynodon dactylon* (Bahama Grass) in Cuba and Jamaica, and *Saccharum officinale* (Sugar Cane) is also a food-plant. On Little Cayman the butterfly was often found near *Sporobolus virginicus* (Seashore Dropseed Grass) (Askew 1980). The brown larva rests in a leaf-roll during daylight, emerging at night to feed; it also pupates in its shelter. The adult butterfly has been observed feeding at flowers of *Argusia gnaphalodes* on Little Cayman.

Reputed Cayman Butterflies

In addition to the butterfly species already covered, some others have claims to be admitted to the list of Cayman Island butterflies. There is, however, a degree of uncertainty about their Caymanian status so that we feel further evidence of their occurrence in the Cayman Islands is needed.

ITHOMIIDAE

Cuban Clearwing
Greta cubana (Herrich-Schäffer, 1862)

Ithomiidae is an almost exclusively South American family, but two species fly in the Greater Antilles. They are weak-flying, shade-loving insects, reluctant to leave the woods and forests in which they breed. *Greta* is included here on the basis of a single sighting, most probably of the Cuban species *G. cubana*, on Grand Cayman. Described by Hernández (2004) as a butterfly of dark mountain woods, *G. cubana* is restricted to three mountainous areas of Cuba, one of which is the southerly Sierra Maestra.

Greta cubana is a very delicate, slender-bodied butterfly with elongated and mostly transparent wings. On the upper surface wing apices and margins are black, and there is an oblique black bar at the end of the forewing cell; on the under surface these parts are reddish brown.

On 28 July 2005, Frank Roulstone (National Trust manager) and Dr John Vlitis noticed a Clearwing Butterfly in the bush just off the road leading to the northern end of Grand Cayman's Mastic Trail. The insect was neither captured nor photographed, but there seems little doubt that it was an ithomiid, and the Cuban Clearwing is by far the likeliest candidate. In view of the sedentary habits of Ithomiidae, voluntary dispersal is improbable. In July 2005 there was the monthly record number of five named tropical storms and hurricanes in the area. Hurricane Dennis travelled north of Jamaica and the Sister Islands, skirting the south and west coasts of Cuba to make landfall there, almost due north of Grand Cayman, on 8 July. It passed within 165 miles of Grand Cayman with wind speeds up to 150 mph. The anticlockwise wind circulation could have carried butterflies to Grand Cayman from Cuba but not from Jamaica, and we believe that the butterfly seen on 28 July was a Cuban Clearwing rather than the Jamaican *Greta diaphana* (Drury).

The slender, smooth larva of *G. cubana* feeds on *Cestrum* (Solanaceae); *C. diurnum* (Day Jessamine) is not uncommon in the Cayman Islands. Pupae are silver coloured and look like small mirrors (DeVries 1987). Adult butterflies are particularly interesting in that males gather together in a lek, and females are attracted to the lek by pheromones produced by the males from chemicals obtained from plants of the family Boraginaceae visited especially for this purpose (Pliske *et al.* 1976, DeVries 1987).

SATYRIDAE
Common Ringlet
Calisto herophile (Hübner, 1823)

The family Satyridae, often considered as a subfamily of Nymphalidae, is well represented in temperate regions, but in the West Indies the only genus is *Calisto*. Typical of Satyridae, these are mainly brown butterflies with small eye-spots, and their larvae feed on grasses. *Calisto*, however, is remarkable for its radiation into forty distinct species. This is centred on the island of Hispaniola where no fewer than thirty-five species have evolved. Two other species are found in Cuba and the Bahamas, and one each in Jamaica, Puerto Rico and Anegada (British Virgin Islands) (Smith *et al.* 1974).

Mounted specimen of Common Ringlet in National Trust donated by Joanne Ross, upper surface above, lower below

Calisto herophile is very common in Cuba and the Isle of Pines, and as a separate subspecies in the Bahamas. The caterpillar feeds on grasses and its life history is described by Dethier (1940). The chances of a live specimen of such a relatively weak-flying, small and sedentary butterfly reaching the Cayman Islands unaided must be remote, and the single individual of *C. herophile* that is believed to have been caught on Grand Cayman must have been an accidental import from Cuba. This specimen, which is in the National Trust reference collection, is unlabelled, but we are informed by Joanne Ross that it was taken *circa* 2000 in the screened porch of her house in south George Town.

NYMPHALIDAE
Common Buckeye
Junonia coenia Hübner, 1822

Brown & Heineman (1972: 179) quote from a letter from C. Bernard Lewis: '...in the collection of the Institute of Jamaica ... amongst specimens collected in the Cayman Islands subsequent to the report published by the Carnegie Museum (Carpenter & Lewis, 1943) there is one typical *coenia*, one very close to typical *coenia* and several possible intergrades.'

Junonia coenia is widespread in North America and its range extends southwards into Central America, Bermuda, Cuba and the Isle of Pines. It could easily stray to the Cayman Islands. *J. coenia* resembles *J. genoveva* more closely than *J. evarete*, having a broad, white forewing fascia. It may be distinguished by this fascia extending down the inner as well as the outer side of the large eye-spot, and by its very large hindwing anterior eye-spot which is somewhat larger than the forewing eye-spot and more than twice the diameter of the hindwing posterior eye-spot.

The caterpillar feeds on *Plantago* species (Plantains), and on species of Scrophulariaceae and Verbenaceae.

PIERIDAE
Small White
Pieris (Artogeia) rapae (Linnaeus, 1758)

This Palaearctic species was introduced into the Nearctic region in the nineteenth century, and has successfully colonized most of Canada and the United States. It is found throughout Florida where it can be a pest of brassicas. The only certain West Indian record, however, is of a single male, thought to have been imported with vegetables, which was found in a green-grocery in Kingston, Jamaica in 1960 (Smith *et al.* 1974). An unlabelled male *Pieris rapae* in the National Trust was deposited by Joanne Ross who thought that it had been caught in her George Town garden *circa* 2001 but the provenance of the specimen is very uncertain.

Bush Sulphur
Pyrisitia dina (Poey, 1832)

A resting specimen of what is possibly this species was photographed by Frank Roulstone on 9 December 2007 on Little Cayman.

P. dina bears a resemblance to *P. nise* (p. 96) but is larger (FWL 16-22 mm.). The male is rich yellow above with an orange flush on the outer quarter of the hindwing and a narrow black border to the forewing only. The female is duller and darker with more orange colouration on the hindwing upper surface.

The Bush Sulphur ranges from the south of Florida (colonized during the 1970s), Mexico to Panama, and in the Bahamas, Cuba, Jamaica and Hispaniola. The species was described from Cuba where 'it is very common throughout the island, flying in fields and gardens in company with [*Abaeis*] *nicippe*, [*P.*] *lisa* and *E. daira*, and in woodland clearings' (Smith *et al.* 1994). In contrast to most other species of *Eurema* and *Pyrisitia*, the Bush Sulphur inhabits wooded or bushy places and tends to avoid open areas (DeVries 1987). A larval food-plant in Jamaica, which also grows in the Cayman Islands, is *Alvaradoa amorphoides* (Wild Spanish Armada) (Simaroubaceae).

Kricogonia species

C. Bernard Lewis, visiting Cayman Brac early in April 1940 saw a battered specimen of an unidentified species of *Kricogonia* and reported that islanders found them abundantly at certain times (Carpenter & Lewis 1943).

Brown & Heineman (1972: 296) report an account by Williams (1930): 'In May and June of 1891, Plaxton (1891) noted large numbers of the genus [*Kricogonia*] migrating in a north-westerly direction in Jamaica and off the Cayman Islands in company with millions of *Phoebis* [*sennae*] *eubule* and *Ascia monuste* and with a smaller number of *Anteos maerula*'.

Two species of *Kricogonia*, *K. lyside* (Godart, 1819) and *K. cabrerai* Ramsden, 1920, fly in Cuba. They are white and creamy yellow butterflies, rather smaller than *Glutophrissa drusilla*, with short antennae not or hardly half as long as the forewing cell and tips of the forewings almost right-angled.

Pink-spot Sulphur
Aphrissa neleis (Boisduval, 1836)

This species was reported (as *Phoebis neleis*) to have been found on Grand Cayman in 1938 (Carpenter & Lewis 1943), and the record is repeated by Schwartz *et al.* (1987), but the single specimen collected, now in the Natural History Museum (London), is a male *Aphrissa statira*, a recognized Caymanian butterfly (p. 105).

PAPILIONIDAE
Machaonides Swallowtail
Heraclides machaonides (Esper, 1796)

D'Abrera (1981) figures a male of this species purported to have been collected on Grand Cayman by the 1938 Oxford University expedition. No reference to the species is made by Carpenter & Lewis (1943) who document the butterflies found in 1938. Collins & Morris (1985) repeat the record, but it is dismissed by Schwartz *et al.* (1987). *H. machaonides* is the commonest Swallowtail on Hispaniola, but is apparently confined to that island (Smith *et al.* 1994).

HESPERIIDAE
Tropical Chequered Skipper
Pyrgus oileus (Linnaeus, 1767)

Dr Eugene J. Gerberg (personal communication) informed us that there is an old literature record of this black and white, chequered Skipper being found on Grand Cayman. We have failed to trace the record. Smith *et al.* (1994) write that *P. oileus* 'does not seem to have reached

the Cayman Islands', although it is entirely feasible that this common West Indian butterfly could turn up here. It is one of the dominant butterflies in Cuba and the Isle of Pines, abundant in gardens, roadsides and open situations generally (Hernández 2004).

Ocola Skipper
Panoquina ocola (Edwards, 1863)

This is a white-spotted, brown Skipper similar to *P. lucas* (p. 127) but smaller (FWL *P. ocola* 16.5 mm., *P. lucas* 17-18 mm.) with reduced white upper surface forewing spots and none in the cell. The hindwing under surface is without the oblique submarginal row of small, whitish spots. Dr Eugene J. Gerberg (personal communication) mentions taking a specimen at Boatswain Point, Grand Cayman on 20 February 1977. This was reported by Askew (1988) and repeated by Smith *et al.* (1994), but confirmation of the occurrence of *P. ocola* in the Cayman Islands is desirable. It is a very easily overlooked butterfly.

Nyctelius Skipper
Nyctelius nyctelius (Latreille, 1824)

Nyctelius nyctelius is tabulated by Riley (1975) as occurring in the Cayman Islands, but no specimen of Caymanian provenance could be found in the Natural History Museum, London where Riley worked. It is found throughout the West Indies 'other than the Bahamas and Cayman Islands' (Smith *et al.* (1994)). *N. nyctelius* resembles *Panoquina lucas* but its hindwing under surface is unspotted and marbled lilac.

BUTTERFLIES
OF THE
CAYMAN ISLANDS

PLATES

PLATE I

Scale x 0.7

Monarch, *Danaus plexippus* (page 27)
1. ♂ ups. Cayman Brac 30.i.2008
2. ♀ ups. Cayman Brac 28.i.2008
    The non-migratory subspecies *D. p. megalippe*.

Queen, *Danaus gilippus* (page 30)
3. ♂ ups. Grand Cayman 16.ii.1992
4. ♀ ups. Grand Cayman 12.viii.1985
5. ♂ unds. Grand Cayman 8.vii.1975
    Narrower black forewing margin and fine white edging to veins on hindwing under surface distinguish the Queen from the Soldier.

Soldier, *Danaus eresimus* (page 34)
6. ♂ ups. Grand Cayman 8.vii.1975
7. ♀ ups. Grand Cayman 12.viii.1985
8. ♂ unds. Grand Cayman 8.vii.1975
    The arc of pale spots on the hindwing is characteristic of the Soldier.

Cuban Red Leaf Butterfly, *Anaea troglodyta* (page 36)
9. ♂ ups. Grand Cayman 21.viii.1985
10. ♀ ups. Grand Cayman 14.iii.1997
11. ♂ unds. Grand Cayman 21.viii.1985

Cayman Brown Leaf Butterfly, *Memphis verticordia* (page 39)
12. ♂ ups. Little Cayman 18.vii.1975
13. ♀ ups. Little Cayman 18.vii.1975
14. ♀ unds. Grand Cayman 18.viii.1985
    Subspecies *M. v. danielana*, endemic to the Cayman Islands. Butterflies on Grand Cayman are larger than those from the Sister Islands.

PLATE II

Scale x 0.7

Antillean Ruddy Daggerwing, *Marpesia eleuchea* (page 41)
1. ♂ ups. Dominican Republic 2.xii.1994

Many-banded Daggerwing, *Marpesia chiron* (page 42)
2. ups. Cuba (Natural History Museum)

Haitian Cracker, *Hamadryas amphichloe* (page 44)
3. ♂ ups. Dominican Republic 29.xi.1994

Mimic, *Hypolimnas misippus* (page 45)
4. ♂ ups. Sulawesi 12.iii.1985

Mangrove Buckeye, *Junonia evarete* (page 46)
5. ♂ ups. Grand Cayman 18.viii.1975
6. ♀ unds. Grand Cayman 29.i.2008

Caribbean Buckeye, *Junonia genoveva* (page 49)
7. ♂ ups. Cayman Brac 30.i.2008
8. ♀ ups. Grand Cayman 9.ii.2006
9. ♀ unds. Grand Cayman 9.ii.2006
The broader, paler band running obliquely across the forewing, the best field character for distinguishing the Caribbean Buckeye from the Mangrove Buckeye, is very apparent in the illustrated specimens but will often be found to be less distinct.

White Peacock, *Anartia jatrophae* (page 51)
10. ♂ ups. Grand Cayman 14.ii.1997
11. ♀ ups. Grand Cayman 12.viii.1985

Crescent Spot, *Phyciodes phaon* (page 55)
12. ♂ ups. Grand Cayman 22.viii.1975
13. ♂ unds. Grand Cayman 12.viii.1975

Painted Lady, *Vanessa cardui* (page 56)
14. ♂ ups. France 29.vii.1977

Malachite, *Siproeta stelenes* (page 53)
15. ♂ ups. Grand Cayman 28.ii.1997
The green colouration fades after death.

Mexican Fritillary, *Euptoieta hegesia* (page 58)
16. ♂ ups. Little Cayman 17.vii.1975
17. ♀ ups. Grand Cayman 7.vii.1975
18. ♂ unds. Little Cayman 14.vii.1975

PLATE III

Scale x 0.7 (figs 1-6)

Gulf Fritillary, *Agraulis vanillae* (page 60)
1. ♂ ups.   Grand Cayman 18.viii.1985
2. ♂ unds. Grand Cayman 18.viii.1985

Zebra, *Heliconius charithonia* (page 65)
3. ♂ ups.   Little Cayman 1.viii.1975

Julia, *Dryas iulia* (page 62)
4. ♂ ups.   Grand Cayman 17.viii.1985
5. ♀ ups.   Grand Cayman 11.x.1995
6. ♀ unds. Grand Cayman 11.x.1995

Scale life size (figs 7-29)

Atala Hairstreak, *Eumaeus atala* (page 67)
7. ♂ ups.   Cayman Brac 29.i.2008
8. ♀ ups.   Cayman Brac 29.i.2008
9. ♂ side view Cayman Brac 29.i.2008

Antillean Hairstreak, *Chlorostrymon maesites* (page 70)
10. ♂ ups.   Jamaica (Natural History Museum)

Cuban Grey Hairstreak, *Strymon martialis* (page 71)
11. ♂ ups.   Little Cayman 17.vii.1975
12. ♂ unds. Little Cayman 19.vii.1975

Drury's Hairstreak, *Strymon acis* (page 73)
13. ♂ ups.   Cayman Brac 9.vii.1975
14. ♂ unds. Little Cayman 5.viii.1975

Dotted Hairstreak, *Strymon istapa* (page 75)
15. ♀ ups.   Grand Cayman 25.viii.1985
16. ♂ ups.   Grand Cayman 18.ii.1997
17. ♂ unds. Little Cayman 24.vii.1975

Fulvous Hairstreak, *Electrostrymon angelia* (page 76)
18. ♂ ups.   Grand Cayman 13.viii.1985

Pygmy Blue, *Brephidium exilis* (page 78)
19. ♀ ups.   Grand Cayman 18.ii.1987
20. ♂ unds. Grand Cayman 21.i.2008

Cassius Blue, *Leptotes cassius* (page 80)
21. ♂ ups.   Grand Cayman 21.ii.1997
22. ♀ ups.   Little Cayman 13.vii.1975
23. ♀ unds. Grand Cayman 6.vii.1975

Hanno Blue, *Hemiargus hanno* (page 82)
24. ♂ ups.   Grand Cayman 18.ii.1997
25. ♀ ups.   Grand Cayman 18.ii.1997
26. ♂ side view Grand Cayman 13.viii.1985

Lucas's Blue, *Cyclargus ammon* (page 84)
27. ♂ ups.   Little Cayman 3.viii.1975
28. ♀ ups.   Grand Cayman 5.vii.1975
29. ♀ side view Grand Cayman 2.i.2008

PLATE IV

Scale x 0.7

Florida White, *Glutophrissa drusilla* (page 86)
  1. ♂ ups. Little Cayman 2.viii.1975
  2. ♀ ups. Little Cayman 2.viii.1975
  3. ♂ unds. Little Cayman 13.vii.1975
  4. ♀ unds. Little Cayman 25.vii.1975

Great Southern White, *Ascia monuste* (page 88)
  5. ♂ ups. Little Cayman 11.vii.1975
  6. ♀ ups. Little Cayman 15.vii.1975
  7. ♂ unds. Little Cayman 15.vii.1975

Giant Brimstone, *Anteos maerula* (page 99)
  8. ♂ ups. Grand Cayman 21.viii.1985
         The first specimen found on land in the Cayman Islands.

Cloudless Sulphur, *Phoebis sennae* (page 100)
  9. ♂ ups. Little Cayman 31.vii.1975
  10 ♀ ups. Grand Cayman 2.viii.1985

Cloudless Orange, *Phoebis agarithe* (page 102)
  11. ♀ ups. Grand Cayman 1.x.1995
         The oblique spot-line across the forewing under surface, just visible also on the upper
         surface, is straight and not dislocated as in the other Cayman species of *Phoebis*.

Orange-barred Sulphur, *Phoebis philea* (page 103)
  12. ♂ ups. Mexico (Natural History Museum)
  13. ♀ ups. Mexico (Natural History Museum)

Migrant Sulphur, *Aphrissa statira* (page 105)
  14. ♂ ups. Grand Cayman 26.i.2006
  15. ♀ ups. (yellow form) Grand Cayman 1.ii.2006
  16. ♀ ups. (white form) Grand Cayman 26.i.2006

Orbed Sulphur, *Aphrissa orbis* (page 106)
  17. ♂ ups. Grand Cayman 25.viii.1998
  18. ♀ ups. (malformed specimen)
         Grand Cayman 11.iii.2004

PLATE V

Scale life size

Barred Sulphur, *Eurema daira* (page 91)
1. ♂ ups. Cayman Brac 8.viii.1975
2. ♀ ups. Little Cayman 2.vii.1975

False Barred Sulphur, *Eurema elathea* (page 92)
3. ♂ ups. Grand Cayman 6.x.1995
4. ♂ ups. (poorly developed sex brand) Grand Cayman 14.ii.1997
5. ♀ ups. Grand Cayman 14.ii.1997
6. ♀ side view (lightly marked specimen) Cayman Brac 26.i.2008
7. ♀ side view (heavily marked specimen) Cayman Brac 26.i.2008

Shy Sulphur, *Pyrisitia messalina* (page 94)
8. ♂ ups. Grand Cayman v or vi.1938
9. ♀ ups. Grand Cayman v or vi.1938
10. ♀ unds. Grand Cayman v or vi.1938
These three specimens were captured by C. B. Lewis and G. H. Thompson during the Oxford University expedition of 1938. They are now in the Natural History Museum, London. The Shy Sulphur has not been seen since in the Cayman Islands.

Little Sulphur, *Pyrisitia lisa* (page 95)
11. ♂ ups. Grand Cayman 18.viii.1985
12. ♀ ups. Cayman Brac 8.viii.1975
13. ♂ unds. Little Cayman 4.viii.1975
14. ♀ unds. Little Cayman 5.viii.1975

Mimosa Sulphur, *Pyrisitia nise* (page 96)
15. ♂ ups. Mexico (Natural History Museum)
16. ♀ ups. Guatemala (Natural History Museum)

Black-bordered Orange, *Abaeis nicippe* (page 97)
17. ♂ ups. Little Cayman 19.vii.1975
18. ♀ ups. Little Cayman 21.vii.1975
19. ♂ unds. Little Cayman 19.vii.1975

Dainty Sulphur, *Nathalis iole* (page 98)
20. ♂ ups. Grand Cayman 22.viii.1985
21. ♀ ups. Little Cayman 2.viii.1975
These are the only specimens known to have been found in the Cayman Islands.

PLATE VI

Scale x 0.7 (figs 1-4)

Gold Rim Swallowtail, *Battus polydamas* (page 108)
    1. ♂ ups.   Grand Cayman 8.ii.2008

Andraemon Swallowtail, *Heraclides andraemon* (page 110)
    2. ♂ ups.   Grand Cayman 15.viii.1985,
                  The large subspecies *H. a tailori* endemic to Grand Cayman
    3. ♂ ups.   Little Cayman 25.i.2008
                  The smaller nominate subspecies found in the Sister Islands

Dusky Swallowtail, *Heraclides aristodemus* (page 113)
    4. ♂ ups.   Little Cayman 17.vii.1975

Scale life size (figs 5-17)

Mangrove Skipper, *Phocides pigmalion* (page 115)
    5. ♂ ups.   Little Cayman 25.vii.1975
    6. ♂ unds.   Little Cayman 25.vii.1975

Long-tailed Skipper, *Urbanus proteus* (page 117)
    7. ♂ ups.   Grand Cayman 14.viii.1975

Dorantes Skipper, *Urbanus dorantes* (page 119)
    8. ♂ ups.   Grand Cayman 17.viii.1985

Three-spotted Skipper, *Cymaenes tripunctus* (page 120)
    9. ♂ ups.   Grand Cayman 18.viii.1985

Fiery Skipper, *Hylephila phyleus* (page 122)
    10. ♂ ups.   Grand Cayman 27.ii.1997
    11. ♀ ups.   Grand Cayman 17.viii.1985

Striped Skipper, *Atalopedes mesogramma* (page 123)
    12. ♂ ups.   Cayman Brac 30.i.2008
    13. ♂ unds.   Cayman Brac 30.i.2008

Monk, *Asbolis capucinus* (page 125)
    14. ♂ ups.   Grand Cayman 5.iii.2004

Sugar Cane Skipper, *Panoquina lucas* (page 127)
    15. ♂ ups.   Grand Cayman 12.viii.1985

Obscure Skipper, *Panoquina panoquinoides* (page 129)
    16. ♂ ups.   Little Cayman 14.vii.1975

Canna Skipper, *Calpodes ethlius* (page 126)
    17. ♀ ups.   Grand Cayman 8.ii.2008

Larval Food-plants and Nectar Flowers

Butterfly caterpillars feed on the leaves or sometimes flower-parts of green plants and the adult insects visit flowers for nectar, and in a few instances pollen. Whilst most butterflies are not very particular about their sources of nectar, taking it where and when available, their selection of plant species on which to lay their eggs is specialized. Most species of butterfly exploit a very limited range of related larval food-plants, and the larvae of only a few have a diet which extends beyond a single plant family.

In this section we tabulate the larval food-plants and nectar sources of Cayman butterflies. With respect to nectar flowers, only those which have been seen to be visited in the Cayman Islands are listed. Our knowledge of larval food-plants in the Cayman Islands is, however, rather limited, and we include in our list plants that have been recorded as larval food elsewhere if they are native or introduced to the Cayman Islands.

Parks and gardens are a very valuable habitat for many species of butterflies, and are likely to become increasingly important as more and more native vegetation is sacrificed to development. The quality of gardens as a resource for Cayman butterflies can be greatly enhanced by the inclusion of native larval food-plants, and we urge gardeners to consider growing some of those listed in the table. We have marked with an asterisk those that we especially recommend. Butterflies are an adornment to any garden and the damage to plants caused by feeding caterpillars ('worms') is a small price to pay for the pleasure that a garden full of butterflies brings. By including larval food-plants in a garden landscape, and avoiding the application of insecticides wherever possible, a valuable contribution to butterfly conservation will be made.

Food-plants [f] and nectar flowers [n] of Cayman butterflies are arranged by their scientific names in families, both in alphabetic order. Cayman common names for the plants are given, and also whether they are native or naturalized [N] in the Cayman Islands, or a cultivated horticultural introduction [C]. Instances where a known food-plant of a Cayman butterfly is present in the islands, but its exploitation is unconfirmed, are indicated by parentheses [(f)].

Acanthaceae

Asystasia gangetica	Bermuda Primrose	N	n	*Danaus plexippus*
			n	*Phoebis sennae*
			n	*Battus polydamas*
Blechum pyramidatum (= *brownei*)	Blechum	N	(f)	*Junonia genoveva*
			(f)	*Anartia jatrophae*
			f?	*Siproeta stelenes*

Ruellia tuberosa	Duppy Gun, Heart Bush	N	(f)	*Junonia genoveva*
			(f)	*Siproeta stelenes*
Aizoaceae				
Sesuvium portulacastrum	Sea-pusley	N	n	*Brephidium exilis*
			n	*Cyclargus ammon*
Amaranthaceae				
Blutaparon vermiculare	Silverhead	N	n	*Brephidium exilis*
			n	*Hemiargus hanno*
Anacardiaceae				
Schinus terebinthifolius	Brazilian Pepper	C	(f)	*Electrostrymon angelia*
Apocynaceae				
*Asclepias curassavica	Red Top	N	f,n	*Danaus plexippus*
			f,n	*Danaus gilippus*
			f,n	*Danaus eresimus*
			n	*Dryas iulia*
			n	*Heraclides andraemon*
*Calotropis gigantea	Giant Milkweed	C	f,n	*Danaus plexippus*
			f	*Danaus gilippus*
Calotropis procera	French Cotton	N,C	f,n	*Danaus plexippus*
			f	*Danaus gilippus*
Metastelma palustre	Gulf Coast Swallowwort	N	(f)	*Danaus gilippus*
			(f)	*Danaus eresimus*
Sarcostemma clausum	White Twinevine	N	f?	*Danaus gilippus*
			(f)	*Danaus eresimus*
Arecaceae (Palmae)				
Chrysalidocarpus lutescens	Areca Palm	C	(f)	*Asbolis capucinus*
Coccothrinax proctorii	Silver Thatch Palm	N	(f)	*Asbolis capucinus*
Cocos nucifera	Coconut Palm	N	(f)	*Asbolis capucinus*
Phoenix dactylifera	Date Palm	C	(f)	*Asbolis capucinus*
Phoenix roebelenii	Pygmy Date Palm	C	(f)	*Asbolis capucinus*
Roystonia regia	Royal Palm	N	(f)	*Asbolis capucinus*
Thrinax radiata	Bull Thatch Palm	N	(f)	*Asbolis capucinus*
Veitchia merrilli	Christmas Palm	C	(f)	*Asbolis capucinus*
Aristolochiaceae				
Aristolochia gigantea	Dutchman's Pipe	C	f	*Battus polydamas*
Aristolochia odoratissima	Fragrant Dutchman's Pipe	N	f	*Battus polydamas*

Asteraceae (Compositae)

Ageratum conyzoides	Wild Ageratum	N	n	*Danaus gilippus*
			n	*Danaus eresimus*
			n	*Atalopedes mesogramma*
Ageratum littorale	Seaside Ageratum	N	n	*Danaus gilippus*
			n	*Danaus eresimus*
Ambrosia hispida	Running Wormwood	N	n	*Hemiargus hanno*
Bidens alba	Spanish Needle	N	n	many butterfly species
Borrichia arborescens	Bay Candlewood, Sea Ox-eye Daisy	N	n	*Danaus gilippus*
			n	*Atalopedes mesogramma*
Chromolaena odorata (= *Eupatorium odoratum*)	Jack-in-the-Bush	N	n	*Dryas iulia*
			n	*Glutophrissa drusilla*
Gynura aurantiaca	Velvet Plant	C	n	*Danaus gilippus*
			n	*Danaus eresimus*
Spilanthes urens	White Button, Pigeon Coop	N	n	*Danaus eresimus*
			n	*Phyciodes phaon*
			n	*Strymon istapa*
			n	*Hemiargus hanno*
			n	*Cyclargus ammon*
			n	*Eurema elathea*
Tridax procumbens	Rabbit Thistle	N	n	*Agraulis vanillae*
			n	*Strymon istapa*
			n	*Cyclargus ammon*

Avicenniaceae

Avicennia germinans	Black Mangrove	N	f	*Junonia evarete*

Bataceae

Batis maritima	Saltwort	N	(f)	*Ascia monuste*

Bignoniaceae

Callichlamys latifolia		C	(f)	*Aphrissa statira*

Boraginaceae

Argusia gnaphalodes	Sea Lavender	N	n	*Strymon martialis*
			n	*Strymon acis*
			n	*Strymon istapa*
			n	*Phocides pigmalion*
			n	*Hylephila phyleus*
			n	*Panoquina lucas*
			n	*Panoquina panoquinoides*

Cordia spp.		N,C	n	*Agraulis vanillae*
			n	*Anteos maerula*
			n	*Phoebis sennae*
			n	*Heraclides aristodemus*
Heliotropium angiospermum	Scorpion Tail	N	n	*Strymon istapa*
Tournefortia volubilis	Aunt Eliza Bush	N	n	*Heraclides aristodemus*
			n	*Urbanus proteus*

Brassicaceae (Cruciferae)

Brassica spp.	Cabbage, Kale	C	(f)	*Ascia monuste*
			(f)	*Urbanus proteus*
Cakile lanceolata	Coastal Sea Rocket	N	f	*Ascia monuste*
Lepidium virginicum	Pepperweed	N	f	*Ascia monuste*

Cannaceae

Canna flaccida	Canna Lily	C	f	*Calpodes ethlius*
Canna indica	Indian Shot	C	(f)	*Calpodes ethlius*

Capparaceae

Capparis cynophallophora	Headache Bush	N	f	*Ascia monuste*
Capparis flexuosa	Bloody Head-Raw Bones	N	(f)	*Glutophrissa drusilla*
			f	*Ascia monuste*
Cleome viscosa	Spider Flower	N	f	*Ascia monuste*

Chenopodiaceae

Salicornia perennis	Glasswort	N	f,n	*Brephidium exilis*

Convolvulaceae

Ipomoea spp.		N	(f)	*Hypolimnas misippus*
			(f)	*Vanessa cardui*

Euphorbiaceae

Croton linearis	Rosemary	N	f	*Anaea troglodyta*
			(f),n	*Strymon acis*
			n	*Ascia monuste*
Croton lucidus	Fire Bush	N	(f)	*Memphis verticordia*
Croton nitens	Wild Cinnamon	N	f	*Memphis verticordia*
Jatropha integerrima	Jatropha	C	n	*Danaus gilippus*
			n	*Danaus eresimus*
			n	*Strymon istapa*
			n	*Ascia monuste*

			n	*Heraclides andraemon*
Jatropha multifida	Coral Plant	C	n	*Anartia jatrophae*
			n	*Euptoieta hegesia*
			n	*Agraulis vanillae*
			n	*Dryas iulia*
			n	*Heliconius charithonia*

Fabaceae (Leguminosae)

Abrus precatorius	Licorice	N	(f)	*Hemiargus hanno*
Bauhinia divaricata	Bull Hoof	N	n	*Danaus eresimus*
			n	*Leptotes cassius*
			n	*Heraclides andraemon*
			n	*Heraclides aristodemus*
			n	*Phocides pigmalion*
Caesalpinia bonduc	Cockspur, Gray Nickel	N	(f)	*Cyclargus ammon*
Caesalpinia pulcherrima	Pride of Barbados	C	n	*Anartia jatrophae*
			n	*Euptoieta hegesia*
			n	*Agraulis vanillae*
			n	*Dryas iulia*
			n	*Heliconius charithonia*
			n	*Anteos maerula*
			n	*Phoebis sennae*
			(f)	*Phoebis philea*
			(f)	*Aphrissa orbis*
			n	*Heraclides andraemon*
			n	*Phocides pigmalion*
Calliandra cubensis	Powderpuff	N	(f)	*Aphrissa statira*
Chamaecrista nictitans	Wild Shame-face	N	(f)	*Hemiargus hanno*
Clitoria ternatea	Bluebell	N	(f)	*Urbanus proteus*
			(f)	*Urbanus dorantes*
Crotalaria spp.	Sweet Pea, Rattlebox	N	(f)	*Hemiargus hanno*
Dalbergia ecastaphyllum	Coin Vine	N	(f)	*Aphrissa statira*
Desmanthus virgatus	Ground Tamarind	N	(f)	*Pyrisitia lisa*
			(f)	*Pyrisitia nise*
Desmodium tortuosum	Twisted Tick-trefoil	N	f	*Urbanus proteus*
			(f)	*Urbanus dorantes*
Galactia striata	Milk Pea	N	f?	*Leptotes cassius*
Gliricidia sepium	Quickstick	N	(f)?	*Anteos maerula*

Haematoxylum campechianum	Logwood	N	n	*Siproeta stelenes*
			n	*Leptotes cassius*
Mimosa pudica	Shame-face	N	(f)	*Hemiargus hanno*
			(f)	*Pyrisitia lisa*
			(f)	*Pyrisitia nise*
Phaseolus lathyroides	Wild Dolly	N	(f)	*Hemiargus hanno*
			(f)	*Urbanus proteus*
			(f)	*Urbanus dorantes*
Piscidia piscipula	Dogwood, Fishpoison Tree	N	(f)	*Electrostrymon angelia*
*Pithecellobium unguis-cati	Catclaw Blackbead, Privet	N	f?	*Phoebis agarithe*
			(f)	*Phoebis philea*
*Senna alata	Candle Bush	C	(f)	*Hemiargus hanno*
			f	*Phoebis sennae*
*Senna occidentalis	Dandelion, Septic Weed	N	f	*Phoebis sennae*
*Senna surattensis	Scrambled Eggs Cassia	C	f	*Phoebis sennae*
Senna and *Cassia* spp.		N,C	(f)	*Eurema elathea*
			(f)	*Pyrisitia messalina*
			(f)	*Pyrisitia lisa*
			(f)	*Pyrisitia nise*
			(f)	*Abaeis nicippe*
			(f)	*Anteos maerula*
			(f)	*Phoebis agarithe*
			(f)	*Phoebis philea*
			(f)	*Aphrissa statira*
Stylosanthes hamata	Donkey Weed	N	f?	*Eurema daira*
			f?	*Eurema elathea*
Vigna luteola	Hairypod Cowpea	N	(f)	*Urbanus proteus*

Lamiaceae (Labiatae)

Salvia spp.	Sages	N	(f)	*Electrostrymon angelia*

Malvaceae

Hibiscus spp.	Hibiscus	N,C	n	*Hamadryas amphichloe*
			n	*Anteos maerula*
			n	*Phoebis sennae*
			n	*Phoebis agarithe*
			n	*Asbolis capucinus*

Moraceae

Artocarpus altilis	Breadfruit	C	(f)	*Marpesia chiron*
			n	*Siproeta stelenes*
Ficus aurea	Wild Fig	N	(f)	*Marpesia eleuchea*
			(f)	*Marpesia chiron*
Ficus carica	Edible Fig	C	(f)	*Marpesia eleuchea*
Maclura (= Chlorophora) tinctoria	Fustic	N	(f)	*Marpesia chiron*

Nyctaginaceae

Bougainvillea sp.	Bougainvillea	C	n	*Danaus plexippus*
			n	*Agraulis vanillae*
			n	*Anteos maerula*
			n	*Phoebis agarithe*
			n	*Urbanus proteus*
Guapira discolor	Cabbage Tree	N	n	*Battus polydamas*

Passifloraceae

Passiflora caerulea	Blue Passionflower	C	f	*Euptoieta hegesia*
			f	*Agraulis vanillae*
Passiflora cupraea	Wild Red Passionflower	N	f	*Euptoieta hegesia*
			f	*Agraulis vanillae*
			f	*Dryas iulia*
Passiflora suberosa	Corky Stem Vine	N	f	*Euptoieta hegesia*
			f	*Agraulis vanillae*
Passiflora spp.	Passionflower Vines	N,C	(f)	*Heliconius charithonia*

Poaceae (Gramineae)

Bambusa vulgaris	Bamboo	C	(f)	*Cymaenes tripunctus*
			(f)	*Panoquina lucas*
Cynodon dactylon	Bermuda Grass	N	(f)	*Hylephila phyleus*
			(f)	*Atalopedes mesogramma*
			(f)	*Panoquina panoquinoides*
Digitaria insularis	Sour Grass	N	(f)	*Hylephila phyleus*
Eleusine indica	Goose Grass	N	(f)	*Atalopedes mesogramma*
Panicum maximum	Guinea Grass	N	(f)	*Cymaenes tripunctus*
			(f)	*Hylephila phyleus*
Paspalum spp.	Crown Grass	N	(f)	*Hylephila phyleus*

			(f)	*Atalopedes mesogramma*
Saccharum officinale	Sugar Cane	N	(f)	*Cymaenes tripunctus*
			(f)	*Hylephila phyleus*
			(f)	*Panoquina lucas*
			(f)	*Panoquina panoquinoides*
Sorghum halepense	Johnson Grass	N	(f)	*Panoquina lucas*
Sporobulus virginicus	Seashore Dropseed Grass	N	(f)	*Panoquina panoquinoides*
Stenotaphrum secundatum	St Augustine Grass	N	(f)	*Hylephila phyleus*

Polygonaceae

Coccoloba uvifera	Sea Grape	N	n	*Danaus eresimus*
			n	*Siproeta stelenes*

Portulacaceae

Portulaca sp.	Purslane	N	(f)	*Hypolimnas misippus*

Rhamnaceae

Colubrina cubensis	Cajon	N	n	*Danaus eresimus*
			n	*Marpesia eleuchea*
			n	*Anartia jatrophae*
			n	*Siproeta stelenes*
			n	*Agraulis vanillae*
			n	*Phoebis sennae*

Rhizophoraceae

Rhizophora mangle	Red Mangrove	N	f,n	*Phocides pigmalion*

Rubiaceae

Morinda royoc	Yellow Root	N	n	*Agraulis vanillae*

Rutaceae

Amyris elemifera	Candlewood	N	f	*Heraclides andraemon*
			(f)	*Heraclides aristodemus*
Citrus aurantifolia	Lime	C	f	*Heraclides andraemon*

Zanthoxylum coriaceum	Shake-hand Tree	N	f	*Heraclides andraemon*
Zanthoxylum fagara	Wild Lime	C	f	*Heraclides andraemon*
			(f)	*Heraclides aristodemus*
Zanthoxylum flavum	Satinwood, Yellow Sanders	N	f	*Heraclides andraemon*

Sapindaceae

Cardiospermum sp.	Balloon Vine	N	(f)	*Chlorostrymon maesites*
Melicoccus bijugatus	Ginep	N	(f)	*Aphrissa statira*

Sapotaceae

Pouteria campechiana	Egg Fruit	C	n	*Siproeta stelenes*

Scrophulariaceae

Bacopa monnieri	Water Hyssop	N	(f)	*Anartia jatrophae*
Stemodia maritima	Seaside Twintip (US)	N	(f)	*Junonia genoveva*

Simaroubaceae

Alvaradoa amorphoides	Wild Spanish Armada	N	(f)	*Pyrisitia dina*

Sterculiaceae

Melochia tomentosa	Velvet-leaf	N	f	*Urbanus proteus*
Suriana maritima	Jennifer, Juniper	N	(f),n	*Strymon martialis*
			(f),n	*Strymon istapa*
			n	*Ascia monuste*
			n	*Hylephila phyleus*
			n	*Atalopedes mesogramma*
Waltheria indica	Buff-coat	N	n	*Agraulis vanillae*
			f	*Strymon istapa*

Turneraceae

Turnera ulmifolia	Dashalong, Catbush	N	f,n	*Euptoieta hegesia*
			n	*Asbolis capucinus*

Ulmaceae

Trema lamarckianum	Trema	N	(f)	*Strymon acis*

156

Verbenaceae

Duranta erecta	Golden Dewdrop	N,C	n	*Danaus eresimus*
			n	*Junonia genoveva*
			n	*Junonia evarete*
			n	*Anartia jatrophae*
			n	*Euptoieta hegesia*
			n	*Agraulis vanillae*
			n	*Phoebis sennae*
Lantana camara	Lantana, Sage	N	n	*Danaus gilippus*
			n	*Hamadryas amphichloe*
			n	*Anartia jatrophae*
			(f),n	*Vanessa cardui*
			n	*Phoebis agarithe*
			n	*Heraclides andraemon*
			n	*Heraclides aristodemus*
			n	*Urbanus proteus*
			n	*Cymaenes tripunctus*
			n	*Hylephila phyleus*
			n	*Asbolis capucinus*
Lippia alba	Providence Mint	C	n	*Anartia jatrophae*
Lippia (= Phyla) nodiflora	Match Head	N	(f),n	*Anartia jatrophae*
			(f),n	*Phyciodes phaon*
			n	*Strymon istapa*
			n	*Cyclargus ammon*
Petitia domingensis	Fiddlewood	N	n	*Memphis verticordia*
			n	*Siproeta stelenes*
Stachytarpheta jamaicensis	Vervine	N,C	n	many butterfly species
Vitex agnus-castus	Butterfly Bush	C	n	*Danaus gilippus*
			n	*Anartia jatrophae*
			n	*Junonia evarete*

Vitaceae

Cissus trifoliata	Sorrel Vine	N	n	*Agraulis vanillae*

Zamiaceae

Zamia integrifolia	Bull Rush, Coontie	N	f	*Eumaeus atala*

Butterfly-like moths

It is not always easy to distinguish butter-flies from moths. Generally speaking, moths are active at night when wing patterns cannot be seen, and they are more sombrely coloured than butterflies. There are, however, some brightly coloured day-flying moths which might be mistaken for butterflies, and to distinguish these some of the characteristic features of moths must be looked for – resting usually with wings held horizontally and not folded vertically over the back, stouter bodies, smaller eyes and, especially, antennae which are thread-like or feathered without a well-defined terminal expansion or club. We mention here just a selection of Cayman moths which might be confused with butterflies.

The most spectacular of Cayman's day-flying moths is undoubtedly **Urania fulgens** (Walker) (Uraniidae), the Green Urania, a very large moth with a wing-span of more than 80 mm. The basal part of the forewing is crossed by six narrow, iridescent blue-green lines, the outermost of which unites with a much broader band on the disc of the wing, distal to which the wing apex is entirely black and there is a long tail on each hindwing. It could very easily be misidentified as a swallow-tail butterfly, but its wing pattern is un-like that of any West Indian swallowtail, and its antennae are typical of those of a moth. *U. fulgens* is not a resident but reaches Grand Cayman from time to time, probably blown off-course during its migrations through Central America. Unlike the butterflies *Marpesia chiron* and *Aphrissa statira*, *Urania* is apparently unable to adjust its line of flight to compensate for wind drift (Srygley *et al.*

1996). Large numbers of *U. fulgens* appeared on Grand Cayman (George Town, Old Man Bay, East End) in October 2005, but the larval food-plant, *Omphalea*, does not seem to grow in the Cayman Islands so that establishment is not a possibility. There is a population in Cuba. *U. boisduvalii* Guérin-Meneville is another Cuban moth which has been found on Grand Cayman, a specimen now in the National Trust being picked up on the beach at Pease Bay in late November 2001 shortly after Hurricane Michelle. *U. boisduvalii* differs from *U. fulgens* in having six rather broader basal stripes on the forewing, the sixth joining a seventh stripe which is not much broader than stripes 2-4, and three or four narrow apical stripes across the wing apex distal to stripe 7. There is a battered, unlabelled specimen of unknown provenance of a third species of *Urania* in the Cayman Islands National Museum. This resembles the South American *U. leilus* (Linnaeus), having a broad band of white marginal spots on the hindwing and a white tail with a narrow black median line.

Ascalapha odorata (Linnaeus) (Noctuidae), the Black Witch or Duppy Bat, is a huge moth and, unlike the others mentioned in this section, it is not brightly patterned, being mostly various shades of brown with dark bands and lines, and having a large but ill-defined eye-spot on each hindwing. It sometimes flies in daylight, although mostly at night, and has a wing-span of up to about 150 mm. At a distance, a day-flying Black Witch could be mistaken for a Gold Rim Swallowtail. *A. odorata* is known from Grand Cayman and Little Cayman. Its larval food-plants include

Cassia and *Pithecellobium* (Fabaceae).

Composia fidelissima (Herrich-Schäffer) (Arctiidae), the Faithful Beauty, is a commonly seen resident day-flying moth with a wing-span of about 50 mm. Marked with red and white spots on iridescent blue-black wings, it is found in the southern United States, Central America, the West Indies and much of South America. Seen flying at a distance on Cayman Brac, it could be mistaken for the Atala Hairstreak butterfly (p. 67). The caterpillar feeds on *Sarcostemma clausum* (Apocynaceae) and both it and the adult moth are probably distasteful to vertebrate predators.

Syntomeida epilais (Walker) (Arctiidae), the Polkadot, is another blue-black moth spotted with white (five spots on each forewing and a pair on the thorax). Its bright orange caterpillars can inflict considerable damage on cultivated Oleander (*Nerium oleander*) (Apocynaceae), a highly poisonous plant, and the conspicuous wing pattern of the adult moth, displayed in slow flight, is almost certainly a warning signal to would-be predators.

Empyreuma affinis Rothschild (Arctiidae), another warningly coloured moth, also has larvae that feed on Oleander. The wings of the moth are red, the forewing with a bluish tint on the veins and outer margin. The stout, bipectinate antennae are black with metallic blue tints and the tips are orange. The body also is metallic blue-black and there are white spots on the thorax and abdomen.

Utetheisa ornatrix bella (Linnaeus) (Arctiidae), the Bella Moth, is sometimes treated as a full species. It is the New World counterpart of the Old World *U. pulchella* (Linnaeus) (Crimson Speckled Footman). A single specimen of *U. o. bella* was found in 1948 on the island of Skokholm off the coast of south-west Wales, but *U. pulchella* is a much more frequent migrant to the British Isles. The larva of the Bella Moth feeds upon species of *Crotalaria* (Fabaceae). The forewing of *U. o. bella* is yellowish white with irregular, transverse white bands, each enclosing a line of small black spots, and the hindwing is pink with marginal black marks.

Aellopos tantalus (Linnaeus) (Sphingidae), Cayman's Hummingbird Hawkmoth, is responsible for many of the reported sightings of hummingbirds in the Cayman Islands. It is a mostly brown-winged moth with a conspicuous silvery white band on the dorsal surface of the second abdominal segment; this is very visible as the moth hovers in front of a flower, wings a blur as it probes for nectar with its long proboscis. Flowers visited include *Bauhinia divaricata*, *Duranta erecta* and *Vitex agnus-castus*. The moth often appears about an hour before sunset and it seems to have a preference for shaded places. It has been observed on all three Cayman Islands. The caterpillar feeds upon Buttonwoods (*Conocarpus* species). Several other species of hawk-moths occur in the Cayman Islands but fly at twilight or are nocturnal.

Melanchroia chephise (Cramer) (Geometridae), Duppy Bush Moth or White-tipped Black, is a small black moth (wingspan about 35 mm.) with narrow white wing-tips. In the Cayman Islands its looper caterpillars are plentiful on *Phyllanthus angustifolius* (Duppy Bush) (Euphorbiaceae) and they also eat *P. acidus* (Chellamella, Otaheite Gooseberry).

Butterflies on Cayman Postage Stamps

Butterflies have featured on the following five issues of postage stamps. Species names are as used in this book, and in a few cases they differ from those appearing on the stamps.

1977	Heliconius charithonia	8c	Danaus gilippus	10c
	Agraulis vanillae	15c	Junonia evarete	20c
	Anartia jatrophae	30c		

[The 5 cents stamp in this set shows the day-flying moth *Composia fidelissima*]

1988	Cyclargus ammon	5c	Phocides pigmalion	25c
	Anaea troglodyta	50c	Heraclides andraemon	$1

1990	Danaus eresimus	5c	Brephidium exilis	25c
	Phyciodes phaon	35c	Agraulis vanillae	$1

1994	Electrostrymon angelia	10c	Eumaeus atala	10c
	Eurema daira	$1	Urbanus dorantes	$1

2005	Danaus gilippus	15c	Euptoieta hegesia	20c
	Siproeta stelenes	25c	Phyciodes phaon	30c
	Phoebis sennae	40c	Heraclides andraemon	90c

References

Ackery, P.R. & Vane-Wright, R.I. 1984. *Milkweed butterflies their cladistics and biology.* British Museum (Natural History), London.

Askew, R.R. 1980. The butterfly (Lepidoptera, Rhopalocera) fauna of the Cayman Islands. *Atoll Research Bulletin* **241**: 121-138.

Askew, R.R. 1988. Butterflies of Grand Cayman, a dynamic island fauna. *Journal of Natural History* **22**: 875-881.

Askew, R.R. 1994. Insects of the Cayman Islands. *In* Brunt, M.A. & Davies, J.E. (eds), *The Cayman Islands: natural history and biogeography*, pp. 333-356. Kluwer Academic Publishers, Dordrecht, The Netherlands.

Beccaloni, G.W., Viloria, Á.L., Hall, S.K. & Robinson, G.S. 2008. *Catalogue of the hostplants of the Neotropical butterflies.* Monografías Tercer Milenio **8**. SEA, Zaragoza & Natural History Museum, London (with RIBES, CYTED & IVIC).

Boppré, M. 1993. The American Monarch: courtship and chemical communication of a peculiar danaine butterfly. *In* Malcolm, S.B. & Zalucki, M. (eds), *Biology and conservation of the Monarch butterfly,* pp. 29-41. Los Angeles County Museum, Los Angeles.

Brown, F.M. & Heineman, B. 1972. *Jamaica and its butterflies.* E.W. Classey, London.

Carpenter, G.D.H. & Lewis, C.B. 1943. A collection of Lepidoptera (Rhopalocera) from the Cayman Islands. *Annals of the Carnegie Museum* **29**: 371-396.

Clench, H.K. 1964. Remarks on the relationships of the butterflies (excluding skippers) of the Cayman Islands. *Occasional Papers on Mollusks, Museum of Comparative Zoology, Harvard* **2**: 381-382.

Collins, N.M. & Morris, M.G. 1985. *Threatened swallowtail butterflies of the world. The IUCN red data book.* IUCN, Gland, Switzerland and Cambridge, U.K.

Comstock, W.P. 1961. *Butterflies of the American tropics: the genus Anaea, Lepidoptera, Nymphalidae.* American Museum of Natural History, New York.

D'Abrera, B. 1981. *Butterflies of the neotropical region. Part 1, Papilionidae and Pieridae.* Lansdowne Editions, East Melbourne.

Dethier, V.G. 1940. Life histories of Cuban Lepidoptera. *Psyche* **47**: 14-26.

DeVries, P.J. 1987. *The butterflies of Costa Rica and their natural history. Papilionidae, Pieridae, Nymphalidae.* Princeton University Press, New Jersey.

Edgar, J.A. 1982. Pyrrolizidine alkaloids sequestered by Solomon Island danaine butterflies. The feeding preferences of the Danainae and Ithomiinae. *Journal of Zoology, London* **196**: 385-399.

Feltwell, J. & Rothschild, M. 1974. Carotenoids in thirty-eight species of Lepidoptera. *Journal of the Zoological Society of London* **174**: 441-465.

Garraway, E. & Bailey, A. 2005. *Butterflies of Jamaica.* Macmillan Education, Oxford.

Gerberg, E.J. & Arnett, R.H. 1989. *Florida butterflies.* Natural Science Publications, Baltimore.

Gilbert, L.E. 1984. The biology of butterfly communities. *In* Vane-Wright, R.I. &

Ackery, P.R. (eds), *The biology of butterflies,* pp. 41-54. Symposium 11, Royal Entomological Society of London. Academic Press, London.

Ground, R.W. 1989. *Creator's Glory.* National Trust for the Cayman Islands, George Town.

Hernández, L.R. 2004. *Field guide of Cuban-West Indies butterflies.* Ediluz, Maracaibo.

Hopf, A.L. 1954. Sex differences observed in larvae of *Danaus berenice. Lepidopterists' News* **8**: 123-124.

Jones, T.H. & Wolcott, G.N. 1922. The caterpillars which eat the leaves of sugar cane in Porto Rico. *Journal of the Department of Agriculture of Porto Rico, San Juan* **6**: 38-50.

Lamas, G. 2004. *Atlas of Neotropical Lepidoptera. Checklist: Part 4A Hesperioidea-Papilionoidea.* Scientific Publishers, Gainesville.

Miller, L.D. & Steinhauser, S.R. 1992. Butterflies of the Cayman Islands, with the description of a new subspecies. *Journal of the Lepidopterists' Society* **46**: 119-127.

Nabokov, V. 1945. Notes on neotropical Plebejinae (Lycaenidae, Lepidoptera). *Psyche* **52**: 1-61.

Nabokov, V. 1948. A new species of *Cyclargus* Nabokov (Lycaenidae, Lepidoptera). *Entomologist* **81**: 273-280.

Nielsen, E.T. 1961. On the habits of the migratory butterfly *Ascia monuste* L. *Biologiske Meddelelser* **23**: 1-81.

Owen, D.F. 1971. *Tropical butterflies.* Clarendon Press, Oxford.

Pierce, N.E. 1984. Amplified species diversity: a case study of an Australian lycaenid butterfly and its attendant ants. *In* Vane-Wright, R.I. & Ackery, P.R. (eds), *The biology of butterflies,* pp. 197-200. Symposium 11, Royal Entomological Society of London. Academic Press, London.

Pliske, T.E. 1973. Factors determining mating frequencies in some New World butterflies and skippers. *Annals of the Entomological Society of America* **66**: 164-169.

Pliske, T.E., Edgar, J.A. & Culvenor, C.C. 1976. The chemical basis of attraction of ithomiine butterflies to plants containing pyrrolizidine alkaloids. *Journal of Chemical Ecology* **2**: 255-262.

Pollard, E. 1977. A method for assessing changes in abundance of butterflies. *Biological Conservation* **12**: 115-134.

Pyle, R.M. 1981. *National Audubon Society field guide to North American butterflies.* Alfred A. Knopf, New York.

Riley, N.D. 1975. *A field guide to the butterflies of the West Indies.* Collins, London.

Rothschild, W. & Jordan, K. 1906. A revision of the American papilios. *Novitates Zoologicae* **13**: 411-752.

Schwartz, A., Gonzalez, F.L. & Henderson, R.M. 1987. New records of butterflies from the West Indies. *Journal of the Lepidopterists' Society* **41**: 145-150.

Scott, J.A. 1972. Biogeography of Antillean butterflies. *Biotropica* **4**: 32-45.

Scott, J.A. 1986a. Distribution of Caribbean butterflies. *Papilio* **3**(n.s.): 1-26.

Scott, J.A. 1986b. *The butterflies of North America. A natural history and field guide.* Stanford University Press.

Scudder, S.H. 1889. *The butterflies of the eastern United States with special reference to New England. Volumes 1-3*, Cambridge, Massachusetts.

Silberglied, R.E. 1977. Communication in the Lepidoptera. *In* Sebeok, T.A. (ed.), *How animals communicate*, pp. 362-402. Indiana University Press, Bloomington.

Silberglied, R.E. 1984. Visual communication and sexual selection among butterflies. *In* Vane-Wright, R.I. & Ackery, P.R. (eds), *The biology of butterflies,* pp. 207-223. Symposium 11, Royal Entomological Society of London. Academic Press, London.

Smith, D.A.S. 1976. Phenotypic diversity, mimicry and natural selection in the African butterfly *Hypolimnas misippus* L. (Lepidoptera: Nymphalidae). *Biological Journal of the Linnaean Society* **8**: 183-204.

Smith, D.S., Miller, L.D. & Miller, J.Y. 1994. *The butterflies of the West Indies and South Florida.* Oxford University Press, Oxford.

Srygley, R.B., Oliveira, E.G. & Dudley, R. 1996. Wind drift compensation, flyways and conservation of diurnal, migrant Neotropical Lepidoptera. *Proceedings of the Royal Society of London B* **263**: 1351-1357.

Stiling, P. 1999. *Butterflies of the Caribbean and Florida.* Macmillan Education Ltd, London and Basingstoke.

Turner, T.W. & Parnell, J.R. 1985. The identification of two species of *Junonia* Hübner (Lepidoptera: Nymphalidae): *J. evarete* and *J. genoveva* in Jamaica. *Journal of Research on the Lepidoptera* **24**: 142-153.

Urquhart, F.A. 1976. *The Monarch butterfly.* University of Toronto Press, Toronto.

Waller, D.A. & Gilbert, L.E. 1982. Roost recruitment and resource utilization: observations on a *Heliconius charitonia* L. roost in Mexico (Nymphalidae). *Journal of the Lepidopterists' Society* **36**: 178-184.

Wetherbee, D.K. 1987. Life stages of *Hamadryas amphichloe diasia* in Hispaniola (Rhopalocera, Nymphalidae). Privately published, Shelburne.

Williams, C.B. 1930. *The migration of butterflies.* Oliver & Boyd, Edinburgh and London.

Witt, T. 1972. Beiträge zur Kenntnis der Gattung *Anaea* Hübner (1819) (Lep., Nymphalidae). *Mitteilungen der Münchner Entomologischen Gesellschaft* **62**: 163-183.

Index to Lepidoptera

Little Cayman

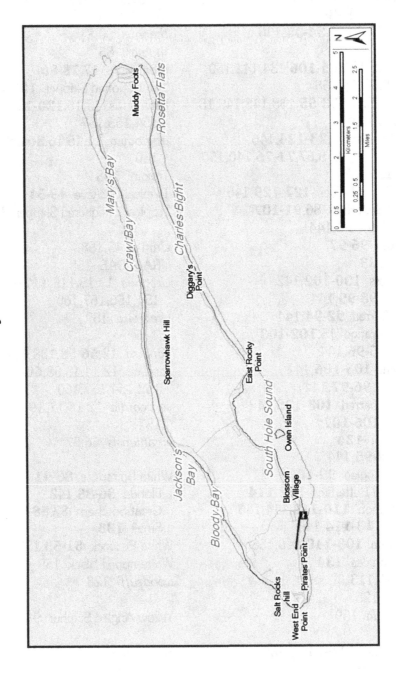

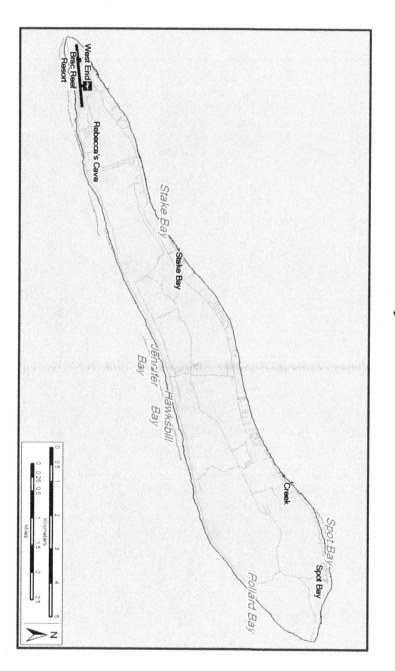

Cayman Brac

Printed in the United States
by Baker & Taylor Publisher Services